Social Networks

Social Networks

Modelling and Analysis

Niyati Aggrawal and Adarsh Anand

CRC Press
Taylor & Francis Group
Boca Raton London New York

CRC Press is an imprint of the
Taylor & Francis Group, an **informa** business

First edition published 2022
by CRC Press
6000 Broken Sound Parkway NW, Suite 300, Boca Raton, FL 33487-2742

and by CRC Press
2 Park Square, Milton Park, Abingdon, Oxon, OX14 4RN

CRC Press is an imprint of Taylor & Francis Group, LLC

Library of Congress Cataloging-in-Publication Data

Names: Aggrawal, Niyati, author. | Anand, Adarsh, author.
Title: Social networks : modelling and analysis / Niyati Aggrawal and Adarsh Anand.
Description: First edition. | Boca Raton : CRC Press, 2022. | Includes bibliographical references and index.
Identifiers: LCCN 2021042172 | ISBN 9780367541392 (hardback) | ISBN 9780367541736 (paperback) | ISBN 9781003088066 (ebook)
Subjects: LCSH: Social media. | Social sciences--Network analysis. | Social media.
Classification: LCC HM742 .A4224 2022 | DDC 302.23/1--dc23/eng/20211021
LC record available at https://lccn.loc.gov/2021042172

ISBN: 9780367541392 (hbk)
ISBN: 9780367541736 (pbk)
ISBN: 9781003088066 (ebk)

DOI: 10.1201/9781003088066

Typeset in Times
by KnowledgeWorks Global Ltd.

Dedication

Dr. Niyati Aggrawal would like to dedicate this book to her father Late Sh. Girish Kumar Aggrawal, her mother Smt. Meenakshi Aggrawal, her husband Sachin and her kids Aarohi and Avik.

Dr. Adarsh Anand would like to dedicate this book to his family and his daughters Ahana and Aavya

Contents

Preface

The objective of this book is to provide essential knowledge of network analysis applicable to the real world. It aims to help academicians, practitioners and researchers understand social media-related concepts in a more refined manner, and to help in understanding the basis of statistical descriptions of networks, analysis of the network structure, various roles and positions of nodes in networks. The book will also provide understanding of the connectivity patterns and link prediction methods. Using various social media platforms such as Facebook, LinkedIn, Twitter and YouTube, this book is an attempt to provide certain mathematical models relevant to understanding the various information diffusion scenarios.

This book is designed to place various social network models that illustrate the analytical reasoning and modelling that can provide both general guidelines to the decision maker as well as quite specific and immediately useful methodical models. Professionals will find this book to be a concise primer on social network analysis from a modelling perspective. The topics covered are organized as follows:

Chapter 1 discusses the basic concepts and terminologies related to network analysis applicable to real-world data, with examples from today's most popular social networks. Chapter 2 discusses the need for scientific analysis wherein the representation of a network as a graph (various kind of graphs) and learnings pertaining to distinguishing weighted, directed and bipartite networks have been discussed.

In Chapter 3, various networking models such as the random graph model, small-world model and scale-free model have been discussed. Centrality measures attempt to find the most central node within a graph. In Chapter 4, various centrality measures for a social media network have been explained.

Chapter 5 associates a relative quantitative assessment with each entity of the network using link-based measures such as PageRank and HITS whereas in Chapter 6, the approaches to link prediction based on measures for analyzing the 'proximity' of vertices in a network have been worked upon. In Chapter 7, various community detection techniques such as node-centric community detection (cliques, *k*-cliques, *k*-clubs) and group-centric community detection have been discussed in detail.

Examining the ego networks of individuals can provide insight into why one individual's perceptions, identity and behaviour differ from another's. In Chapter 8, the focus has been on the individual actor, rather than the network as a whole. The fundamental idea here is the different ways in which individuals are attached to macro-structures. In Chapter 9, the idea of the amount of 'embedding' in whole networks, the extent to which actors find themselves in social structures characterized by dense, reciprocal, transitive and strong ties, has been presented in detail.

Chapter 10 gives an overview of the information diffusion process on social networking platforms and various mathematical modelling frameworks. The four main types of information diffusion – herd behaviour, information cascades, diffusion of innovations and epidemics – have been discussed in detail.

Chapter 11 gives a brief overview about security and privacy issues on social networks, and Chapter 12 provides an overview of the various social networking tools such as Gephi, UCINET, NetworkX, Pajek and many more.

Dr. Niyati Aggrawal
JIIT, Noida, UP, India

Dr. Adarsh Anand
University of Delhi, India

Acknowledgements

We are grateful to various colleagues both in academia and industry who read through draft manuscripts. Sincere thanks to Cindy and Erin, who have guided us throughout the publication process. In addition, we are honoured to be associated with Prof. Mangey Ram, who showed faith in us for carrying out the book compilation work under his series.

Nobody can successfully complete any task without the support of the family. So, our beloved appreciation to our respective families for affording us their patience so that our thoughts could be translated and compiled in the form of this book.

Gratitude also goes to Mr. Sachin Goel, Territory Manager, HDFC Life, and Dr. Deepti Aggrawal, Assistant Professor at USME, DTU, who have helped in articulation of the work by providing critical comments on a time-to-time basis.

Dr. Niyati Aggrawal

Dr. Adarsh Anand

Authors

Dr. Niyati Aggrawal is an accomplished and integrity-driven academic professional with over 10 years of rich experience in varied fields of social network analysis, social media mining, information diffusion on social media, web mining techniques and information security. She has guided various graduate projects in the fields of social network analysis and data mining. She has published several research papers in many reputed international journals and conferences related to the above-mentioned research areas. She has actively participated in various conferences, faculty development programs and workshops.

Dr. Adarsh Anand completed his doctorate in the area of Innovation Diffusion Modelling in Marketing and Software Reliability Assessment. Presently, he is working as an assistant professor in the Department of Operational Research, University of Delhi (India). In 2012, he was conferred with the Young Promising Researcher in the field of Technology Management and Software Reliability by Society for Reliability Engineering, Quality and Operations Management (SREQOM). He is a lifetime member of SREQOM. He is also on the editorial board of the *International Journal of System Assurance and Engineering Management* (Springer). He has guest edited several special issues for journals of international repute. He has edited two books, *System Reliability Management (Solutions and Technologies)* and *Recent Advancements in Software Reliability Assurance,* under the banner of Taylor and Francis (CRC Press). He has publications in journals of national and international repute. His research interests include modelling innovation adoption and successive generations in marketing, software reliability growth modelling and social media analysis.

1 Introduction to Social Networks

In this chapter, an overview of social network properties, design principles, models, algorithms and evolution of networks is discussed.

1.1 CONCEPT OF COMPLEX NETWORKS

In this digital era, social networking has become a basic need and the most preferable option for internet users (netizens) to interact with each other. Gone are the days of disseminating important information through personal interaction or broadcasting, because with the help of social networking platforms, information can be sent all over the world within seconds. People share each and every moment of their day-to-day life with their friends and relatives, even with the world through social networking platforms (Facebook, Twitter, Flicker, Instagram, YouTube and many more). A very large quantity of the data and information is being spread and shared every second using these social platforms. These moments can be in the form of text, audio, video, image, rumours, politics or news. Sharing thoughts and ideas on these platforms have brought people with similar curiosity under one canopy regardless of their human ecology. After viewing the posted content, people show their engagement towards the posted content by hitting like, dislike, share, comment, subscribe, hide and block buttons provided on social networking platforms. These engagements and interests towards any posted content play a vital role in virality. Thus, social networking platforms play a key role in diffusing (making viral) any information on the social web [1]. Social media research areas and domains are outlined in Figure 1.1.

1.2 OVERVIEW OF SOCIAL NETWORK ANALYSIS

1.2.1 Social Networks and Social Networking

A **social network** is a social structure made up of a set of social actors (such as individuals or organizations) and a set of the dyadic ties between these actors. The social network perspective provides a set of methods for analyzing the structure of whole social entities as well as a variety of theories explaining the patterns observed in these structures. The study of these structures uses social network analysis to identify local and global patterns, locate influential entities and examine network dynamics [2].

According to Boyd and Ellison, an online social network site is a web-based service that allows individuals to (1) construct a public or semi-public profile within a bounded system, (2) articulate a list of other users with whom they share a connection and (3) view and traverse their list of connections and those made by others within the system.

DOI: 10.1201/9781003088066-1

Social Media Research Domains

- Social Network
  - Facebook
- Video Social Media:
  - YouTube
  - DailyMotions
- Image Social Media:
  - Instagram
  - Snapchat
  - Flickr
- Research /Professional Social Media
  - DBLP
  - Research Gate
  - Google Scholar
  - LinkedIn
- Micro-Blogging Media
  - Twitter
- Content Curation
  - Pinterest
  - Flipboard

Social Media Research Areas

- Statistical Analysis on Social media
- Social Network Modeling
- Social Network Visualization
- Link Prediction
- Community Detection
- Information Diffusion
- Ego Network
- Security Issues
- Network Motif

FIGURE 1.1 Social network research outline [1]

'Networking' is an activity or process that has always been with us. Simply put, it is the act of reaching out and connecting with others. In this way, we have always been networking – sharing information between individuals that provides enough value to justify maintaining the relationship. Some of these behaviours are purely utilitarian (meaning there is some gain to be had), whereas other behaviours are more focused on the socializing aspects of the relationship itself.

Social network analysis (SNA) is a core aspect of analyzing social networks today. In addition to the usual statistical techniques for data analysis, these networks are investigated using SNA measures. It helps in understanding the dependencies between social entities in the data, characterizing their behaviours and their effect on the network as a whole and over time. Therefore, this chapter attempts to provide a brief overview of SNA in diverse topological networks (static, temporal and

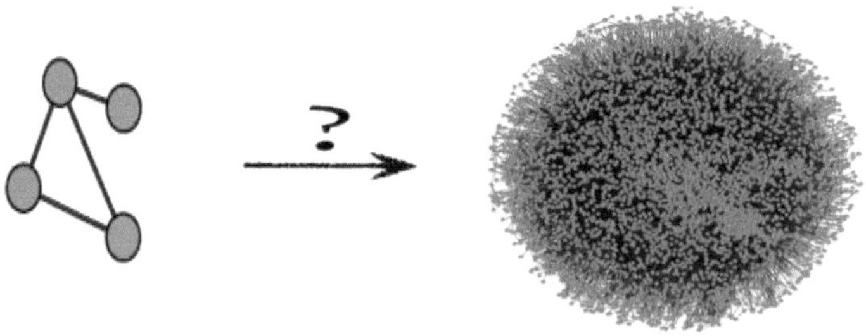

Earlier view of the network   Today's view of the network

FIGURE 1.2 Growth model

evolving networks) and perspective (ego networks). Common tasks of SNA involve the identification of the most influential, prestigious or central actors, using statistical measures; the identification of hubs and authorities, using link analysis algorithms; discovering communities, using community detection techniques and how information propagates in the network, using diffusion algorithms. These tasks are extremely useful in the process of extracting knowledge from networks and, consequently, in the process of problem-solving.

1.2.2 Social Network Visualization and Statistical Analysis

In the real world, a large number of applications can be represented as complex networks, such as social networks (Facebook, Twitter, LinkedIn, co-authorship network etc.), terrorist networks, biological networks (protein interaction networks, metabolic networks etc.), organizational networks, knowledge networks (Wikipedia) and brain and technology networks (electricity networks, the internet etc.). To understand these complex networks, graphical visualization plays a very important role.

Graph is a mathematical representation of a network and it represents the relationship between nodes and edges.

Visualization of the sociogram of the DBLP co-authorship network is depicted in Figure 1.3. Node size denotes the popularity of the author (i.e., a big node size means the author is a popular author). The DBLP data set is represented as an undirected graph and contains 17,280 nodes and 58,539 edges (In this graph, nodes are taken as author and edges connect two nodes/authors according to their association in considered research papers' data set. There will be more than one

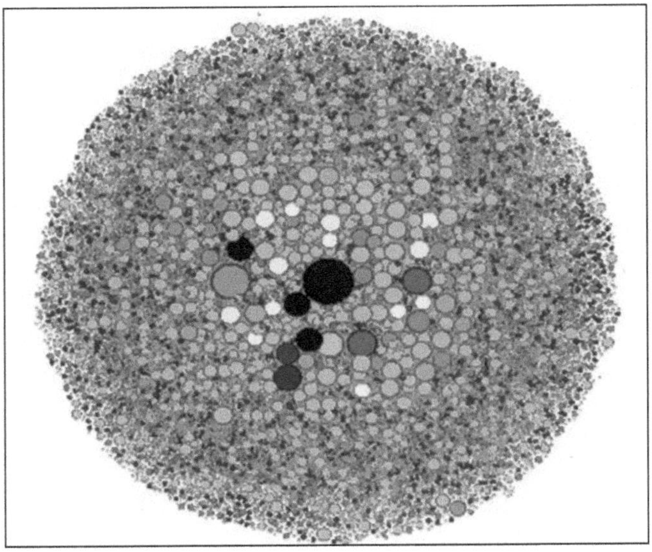

FIGURE 1.3 Visualization of the co-authorship network [3]

TABLE 1.1
Co-authorship Network Data Set [3]

Category	Co-authorship Network
Source of Data Set	http://konect.cc/networks/com-dblp/
Node	Author
Edge type	Co-authorship
Network Format	Undirected
Edge weights	Unweighted
Size	317,080 vertices (authors)
Volume	1,049,866 edges (co-authorships)
Average degree	6.6221 edges/vertex

edge in between two nodes if the researcher co-authored more than one paper. Statistics-based analysis has been done with some varying measuring factors such as connected nodes, clustering coefficient, page rank, centrality of the network, average path length and graph density. Community detection is also done and, thereafter, the graph represents co-authorship network communities. We discuss in detail about collecting, analyzing and visualizing co-authorship data that is further used for understanding the academic collaboration and finding research communities for a specific knowledge domain. Table 1.1 gives information about the co-authorship data set [4].

Statistical measurements such as number of unique authors and co-authors, average degree, connected components, graph density, diameter, path length and clustering coefficient are given in Table 1.2.

TABLE 1.2
Statistical Analysis of Co-authorship Network Result [3]

Description	Statistics
Number of unique authors	17,280
Number of unique co-authors	58,539
Repeat relationships	6756
Average degree	3.38
Connected components	3,713
Vertices in largest component	2,047
Graph density	0.56
Network diameter	38
Average path length	14.383
Modularity	0.994
Average clustering coefficient	0.901

1.2.3 Social Network Modelling

Social networks can be represented and measured by using two basic mathematical models: the random graph model and the scale-free graph model.

Random Graph Model

Various models of random graphs have been proposed for the social network such as the Erdos–Renyi model, small-world model (SWM), preferential attachment model and forest-fire model. Many researchers worked on these basic models and proposed their own models. In mathematics, with the help probability distributions, random graphs can be described.

Scale-Free Graph

A network or graph is known as a scale-free network whose degree distribution follows a power law (richer is getting richer), at least asymptotically. Examples of scale-free networks are the Barabási Albert model (BAM) and the Bianconi–Barabási model (BBM).

1.2.4 Link Prediction

Link prediction algorithms are used to predict the future social network on the basis of the available links in the network. In other words, link prediction algorithms are utilized to predict the likelihood of formation of new connections between any two persons currently not connected with anyone. Real-world examples of link prediction are friend and follower suggestions on social networking platforms, suggestions of the products to customers on e-commerce platforms, suggestions to authors or researchers to work together based on their areas of interest and suggestions to organizations to work with other organizations in their network [5].

1.2.5 Community Detection

Community detection is a very popular research area under social networking domains. It is a process of discovering the cohesive groups or clusters in the network on the basis of similar interests or similar actions on social media. Modularity maximization is the most popular algorithm to discover communities in social networks. Detection of communities on social networking platforms saves time and money in advertising or in spreading a message or product. The detection of communities can be useful in many applications such as multi-casting a message of interest to a community instead of sending it individually to each member in the community or recommendation of a product to a community with a similar interest [1].

1.2.6 Ego Network

An ego network is the network of an individual person; for example, the network of Mr. Amitabh Bachchan on Twitter. Thus, a network has as many egos as it has vertices. Egos can be an individual person, communities, organizations or whole societies. These networks are used to analyze and understand the behaviour of individuals [1].

1.2.7 Network Motifs

Network motifs are basically the patterns that are formed during the connections or interactions among persons in social networks. Study and analysis of the patterns that are formed during the interconnections among social media users provide an attractive research motivation in understanding the complex networks. These patterns are very helpful in dealing with the challenge of studying the connections between structure and type of interaction of social networks [1].

1.2.8 Security and Privacy Issues

A secure platform needs to fulfil the requirements of confidentiality and integrity of user's data. Social media is an inherently open platform and cannot be fully trusted due to the lack of central authentication and authorization mechanisms. Moreover, visibility of shared data is many times beyond the control of the user, thus causing privacy concerns. Social media platforms inadvertently share user's data with third parties and there is a dearth of techniques to control this [3].

Therefore, in this book, we will cover most of these discussed terms in detail.

1.3 SOCIAL MEDIA CONTENT

Research on social network platforms is done on its content posted by diverse medium. In a social media context, that content is also known as a post. This posted content can be self-created text, audio, video, images, news, blogs or shared content by other sources from the internet. Thus, to do research on any social media platform, researchers majorly worked on three main measuring parameters: content characteristics, content dynamics and user characteristics [1].

1.3.1 Content Characteristics

- **YouTube:** Video Length, Video Age, Time of Upload, Title
- **Amazon:** Id, ProductName, Description, Type, Brand, Price, Color, Product Category
- **Twitter:** Tweet Id, Tweet userId, Tweet text, Tweet created at, tweet geo, tweet contributors, tweet coordinates etc.

1.3.2 Content Dynamics

- **YouTube:** View Count, Like Count, Dislike Count, Comment Count, Share Count
- **Amazon:** Review, review rating, time stamp of review, time of review posted
- **Twitter:** Tweet favorited, Tweet in_reply_to_screen_name, Tweet in_reply_to_status_id, Tweet in_reply_to_user_id, Tweet place, Tweet retweeted, Tweet retweet_count, Tweet source, Tweet truncated etc.

1.3.3 User Characteristics

- **YouTube:** Use Id, User name, User Age
- **Amazon:** User name, User screen name, user ID
- **Twitter:** User name, User screen name, user ID, User Created At

1.4 LEVELS OF NETWORK ANALYSIS [6]

In general, social networks are self-organizing, emergent and complex, such that a globally coherent pattern appears from the local interaction of the elements that make up the system. These patterns become more apparent as the network size increases. However, in a global network analysis, social networks are analyzed at the scale relevant to the researcher's theoretical question. Although levels of analysis are not necessarily mutually exclusive, there are three general levels into which networks may fall: micro-level, meso-level and macro-level.

1.4.1 Micro-Level

At the micro-level, social network research typically begins with an individual, growing as social relationships are traced, or may begin with a small group of individuals in a particular social context.

- **Dyadic level:** A dyad is a social relationship between two individuals. Network research on dyads may concentrate on structure of the relationship (e.g., multiplexity, strength), social equality and tendencies towards reciprocity/mutuality.
- **Triadic level:** Add one individual to a dyad, and you have a triad. Research at this level may concentrate on factors such as balance and transitivity, as well as social equality and tendencies towards reciprocity/mutuality.
- **Actor leve:l** The smallest unit of analysis in a social network is an individual in their social setting (i.e., an 'actor' or 'ego'). Ego network analysis focuses on network characteristics such as size, relationship

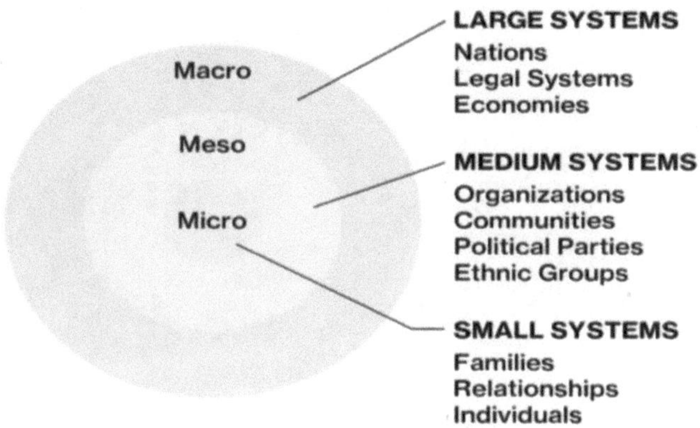

FIGURE 1.4 Levels of network analysis [7]

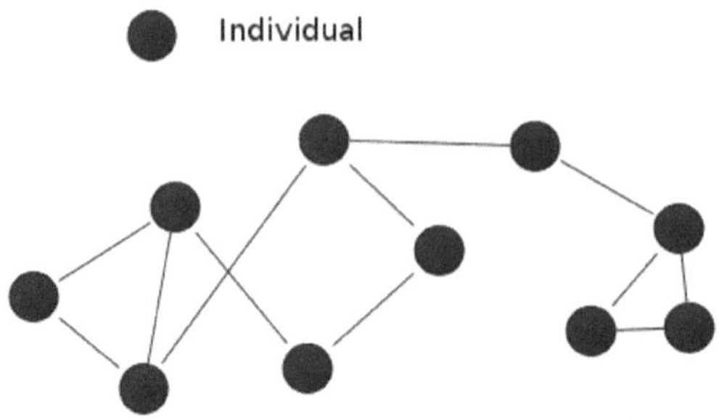

FIGURE 1.5 Social network at the micro-level [6]

strength, density, centrality, prestige and roles such as isolates, liaisons and bridges. Such analyses are most commonly used in the fields of psychology or social psychology, ethnographic kinship analysis or other genealogical studies of relationships between individuals.

- **Subset level:** Subset levels of network research problems begin at the micro-level, but may cross over into the meso-level of analysis. Subset level research may focus on distance and reachability, cliques, cohesive subgroups or other group actions or behaviour.

1.4.2 Meso-Level

In general, meso-level theories begin with a population size that falls between the micro- and macro-levels. However, the meso-level may also refer to analyses that are specifically designed to reveal connections between micro- and macro-levels. Meso-level networks are low density and may exhibit causal processes distinct from interpersonal micro-level networks.

- **Organizations:** Formal organizations are social groups that distribute tasks for a collective goal. Network research on organizations may focus on either intra-organizational or inter-organizational ties in terms

FIGURE 1.6 Social network at the meso-level [6]

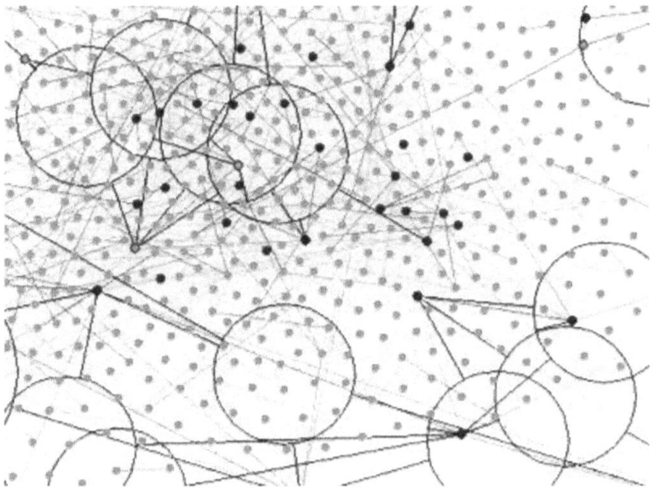

FIGURE 1.7 Social network at the macro-level [6]

of formal or informal relationships. Intra-organizational networks themselves often contain multiple levels of analysis, especially in larger organizations with multiple branches, franchises or semi-autonomous departments. In these cases, research is often conducted at a workgroup level and organization level, focusing on the interplay between the two structures.

1.4.3 Macro-Level

Rather than tracing interpersonal interactions, macro-level analyses generally trace the outcomes of interactions, such as economic or other resource transfer interactions over a large population.

- **Large-scale networks:** Large-scale networks is a term somewhat synonymous with 'macro-level' as used, primarily, in social and behavioural sciences and economics Originally, the term was used extensively in the computer sciences.

1.5 COMPLEX NETWORKS

Larger social networks display features of social complexity, which involves substantial non-trivial features of network topology, with patterns of complex connections between elements that are neither purely regular nor purely random, as do biological and technological networks. Such complex network features include a heavy tail in the degree distribution, a high clustering coefficient, assortative or disassortative among vertices, community structure and hierarchical structure. In the case of agency-directed networks, these features also include reciprocity, triad significance profile (TSP) and other features. In contrast, many of the mathematical models of networks that have been studied in the past, such as lattices and random graphs, do not show these features.

- **Randomly distributed networks:** Exponential random graph models of social networks became state-of-the-art methods of social network analysis in the 1980s. This framework has the capacity to represent social-structural effects commonly observed in many human social networks, including general degree-based structural effects commonly observed in many human social networks as well as reciprocity and transitivity, and at the node-level, homophily and attribute-based activity and popularity effects, as derived from explicit hypotheses about dependencies among network ties. Parameters are given in terms of the prevalence of small subgraph configurations in the network and can be interpreted as describing the combinations of local social processes from which a given network emerges. These probability models for networks on a given set of actors allow generalization beyond the restrictive dyadic independence assumption of micro-networks, allowing models to be built from theoretical structural foundations of social behaviour.
- **Scale-free networks:** A scale-free network is a network whose degree distribution follows a power law, at least asymptotically. In network theory, a scale-free ideal network is a random network with a degree distribution that loosens the size distribution of social groups. Specific characteristics of scale-free networks vary with the theories and analytical tools used to create them; however, in general, scale-free networks have some common characteristics. One notable characteristic in a scale-free network is the relative commonness of vertices with a degree that greatly exceeds the average. The highest-degree nodes are often called 'hubs', and may serve specific purposes in their networks, although this depends greatly on the social context. Another general characteristic of scale-free networks is the clustering coefficient distribution, which decreases as the node degree increases. This distribution also follows a power law. The BAM model of network evolution is an example of a scale-free network.

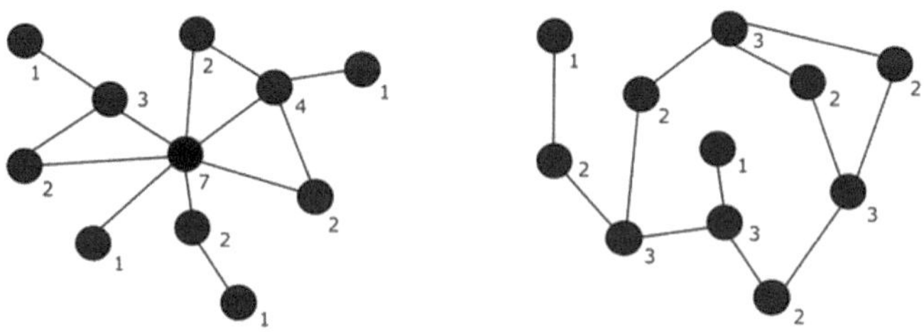

FIGURE 1.8 Random network and scale-free network [8] Graphical representation of a scale-free vs. a random network. The degree (number of interactions) of each node is indicated by the digit below the node. In a scale-free network the majority of nodes interact with just a few other nodes , while only some interact with many other nodes and represent the network hubs (node with degree 7). In a random network all nodes are connected with a uniform probability, resulting in most nodes having the same number of connections. [9]

Therefore, after studying social networks, we will understand patterns and statistical properties of network data, design principles and network models, why networks are organized the way they are, predict behaviour of networked systems and algorithms for analyzing graphs. We will also be able to measure and qualify networks and get the processes about the spread of information in networks.

1.6 PROBLEMS FOR SELF-ASSESSMENT

Q1. What is a social network/social media? List all social networking sites that you know of.

Q2. List all the social media websites that you can think of that may be turned into a social network in the future.

Q3. Explain what features they share? How are they different? Group the chosen social media sites according to their features.

Q4. What are some ways that make social network analysis useful?

Q5. What sites do you personally use? What are the overlapping features (such as using Instagram as a photosharing website in addition to Facebook, which also allows you to share photos)?

Q6. Why do you use two sites instead of one? Do you use different social media sites for different purposes if sites use overlapping features? Why or why not?

REFERENCES

1. Aggrawal, N. *Information diffusion framework for virality prediction of social media content*, PhD thesis, Jaypee Institute of Information Technology.
2. Geethadevi, M. Y., & Kalaimathi, D. H. (2019). A study of academic achievement of B. Ed. teacher trainees with gender and locality. *Journal of Education and Development*, 9(18).
3. Aggrawal N., & Arora A. Visualization, analysis and structural pattern infusion of DBLP co-authorship network using Gephi, *2016 2nd International Conference on Next Generation Computing Technologies (NGCT)*, Dehradun, India, 2016, pp. 494–500, doi: 10.1109/NGCT.2016.7877466
4. Tabassum, S., Pereira, F. S., Fernandes, S., & Gama, J. (2018). Social network analysis: An overview. *Wiley Interdisciplinary Reviews: Data Mining and Knowledge Discovery*, 8(5), e1256.
5. Liang Y., Huang L., Wang Z. (2017). Link prediction in social network based on local information and attributes of nodes. In: *Journal of Physics: Conference Series*, vol. 887 (p. 012043). IOP Publishing.
6. Levels of analysis. https://sites.google.com/site/socialnetworkingbywhitechicks/grimvarna
7. Course Hero. Macro-level, meso-level, and micro-level analysis. https://www.coursehero.com/sg/introduction-to-sociology/macro-level-meso-level-and-micro-level-analysis/
8. Wikipedia. Hub (network science). https://en.wikipedia.org/wiki/Hub_(network_science)
9. Vandereyken, K., Van Leene, J., De Coninck, B., & Cammue, B. (2018). Hub protein controversy: taking a closer look at plant stress response hubs. Frontiers in plant science, 9, 694.

2 Network Statistics and Related Concepts

In this chapter, the representation of a network as a graph and introduction of the elementary characteristics of networks from degree-to-degree distributions, paths to distances and learnings pertaining to distinguishing weighted, directed and bipartite networks have been discussed.

2.1 NETWORKS AND GRAPHS

Networks/graphs are a collection of objects where some pairs of objects are connected by links.

A **network** often refers to real systems such as the web, social networks and metabolic networks, and the language that is used for this is: network, node, link. A **graph** is a mathematical representation of a network web graph or social graph (a Facebook term), and the language that is used for this is: graph, vertex, edge. We will try to make this distinction whenever it is appropriate, but in most cases, we will use the two terms interchangeably.

Components of a network:

Objects: nodes, vertices N
Interactions: links, edges E
System: network, graph $G(N, E)$

Yet, there is a subtle distinction between the two terminologies: the {*network*, *node*, *link*} combination often refers to real systems. The World Wide Web (WWW) is a network of web documents linked by Uniform Resource Locators (URLs); society is

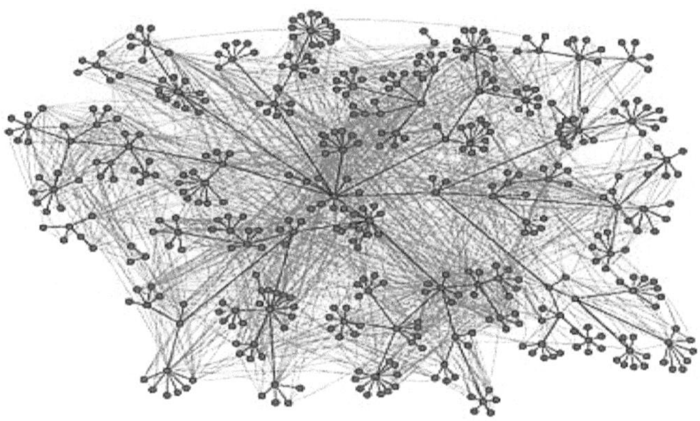

FIGURE 2.1 Network

DOI: 10.1201/9781003088066-2

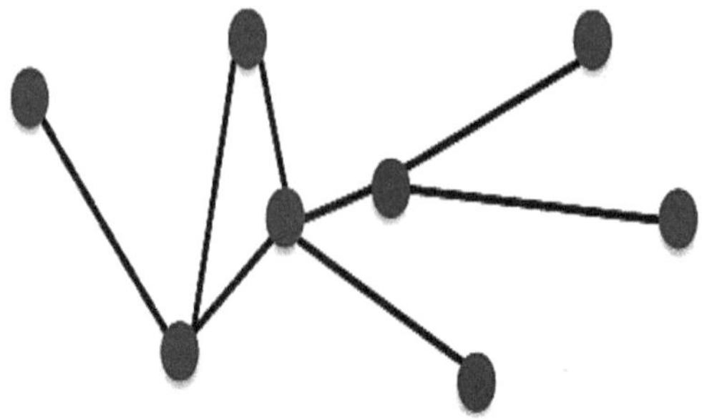

FIGURE 2.2 Network/graph

a network of individuals linked by family, friendship or professional ties; the metabolic network is the sum of all chemical reactions that take place in a cell.

In contrast, we use the terms {*graph*, *vertex*, *edge*} when we discuss the mathematical representation of these networks; we talk about the web graph, the social graph (a term made popular by Facebook) or the metabolic graph. Yet, this distinction is rarely made, so these two terminologies are often synonyms of each other.

If we want to understand a complex system, we first need to know how its components interact with each other. In other words, we need to understand the mathematics behind it. A network is a catalogue of a system's components often called nodes or vertices and the direct interactions between them, called links or edges. This network representation offers a common language to study systems that may differ greatly in nature, appearance or scope. Indeed, as shown in Figure 2.3, three rather different systems have exactly the same network representation.

Figure 2.3 introduces two basic network parameters: **Number of nodes, or N**, represents the number of components in the system. We will often call N the size of the network. To distinguish the nodes, we label them with $i = 1, 2, \ldots, N$. **Number of**

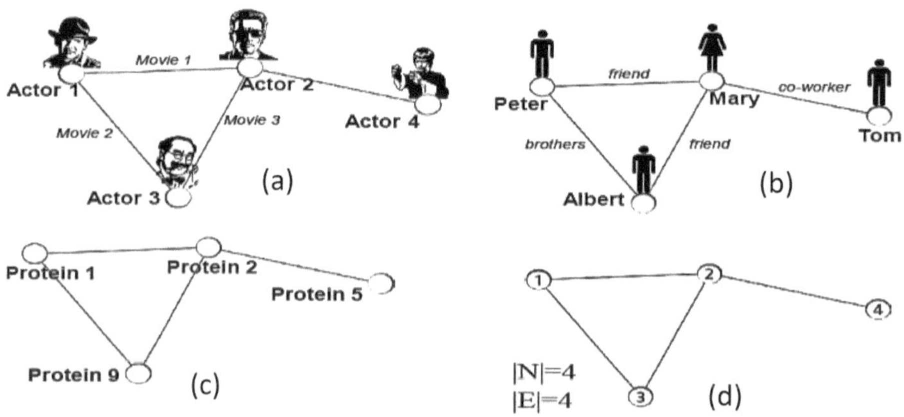

FIGURE 2.3 Three different networks with the same graph [1]

links, which we denote with L, represents the total number of interactions between the nodes. Links are rarely labelled because they can be identified through the nodes that they connect.

For example, the (2, 4) link connects nodes 2 and 4. The networks shown in Figure 2.3 have $N = 4$ *and* $L = 4$.

Different Networks, Same Graph: Figure 2.3 shows a small subset of (a) the Hollywood actor network, where two actors are connected if they played in the same movie; (b) the relationship network, where nodes (individual person) are connected to each other if they are having some kind of relationship with each other and (c) a protein–protein interaction network, where two proteins are connected if there is experimental evidence that they can bind to each other in the cell. While the nature of the nodes and the links differs, these networks have the same graph representation, consisting of $N = 4$ nodes and $L = 4$ links, shown in (d).

2.2 DIFFERENT TYPES OF NETWORKS

The links of a network can be directed or undirected. Some systems have directed links, like the WWW, whose URLs point from one web document to the other, or phone calls, where one person calls the other. Other systems have undirected links, like romantic ties: if I date Janet, Janet also dates me, or like transmission lines on the power grid, on which the electric current can flow in both directions.

2.2.1 Undirected Networks

A network is called undirected if all of its links are undirected (symmetrical, reciprocal). Examples: collaborations, friendship on Facebook, internet, power grid, science collaboration networks.

Connectivity of Undirected Graphs

- **Connected (undirected) graph:** Any two vertices can be joined by a path.
- **A disconnected graph** is made up by two or more connected components.

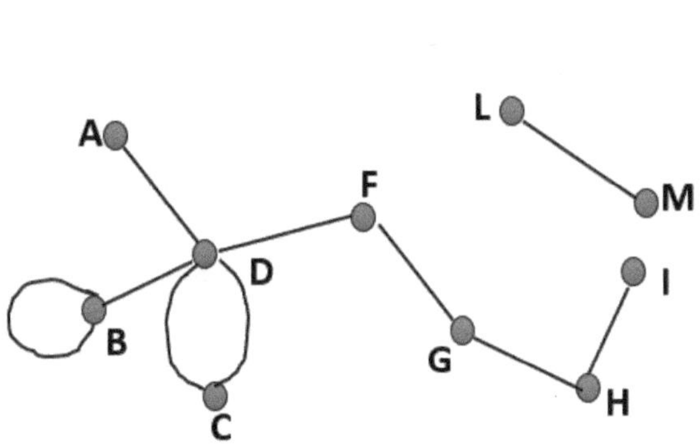

FIGURE 2.4 Undirected network

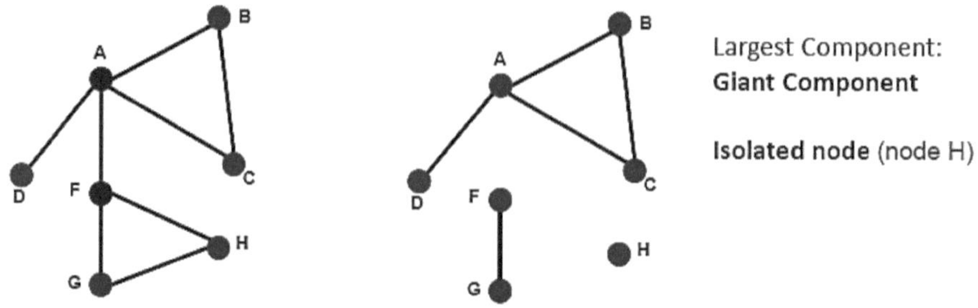

FIGURE 2.5 Connectivitiy in undirected graphs

Bridge Edge

If we erase it, the graph becomes disconnected. Link AF is the bridge edge in Figure 2.5.

Articulation Point

A vertex in an undirected connected graph is an articulation point (or cut vertex) if removing it (and edges through it) disconnects the graph.

2.2.2 Directed Networks

A network is called directed (or digraph) if all of its links are directed (arcs). Examples are phone calls, following on Twitter, WWW, mobile phone calls, citation network.

Some networks simultaneously have directed and undirected links. For example, in the metabolic network, some reactions are reversible (i.e., bidirectional or undirected) and others are irreversible, taking place in only one direction (directed). The choices that we make when we represent a system as a network will determine our ability to use network science successfully to solve a particular problem. For example, the way that we define the links between two individuals dictates the nature of the questions that we can explore:

- By connecting individuals that regularly interact with each other in the context of their work, we obtain the organizational or professional network that plays a key role in the success of a company or an institution and is of major interest to organizational research.

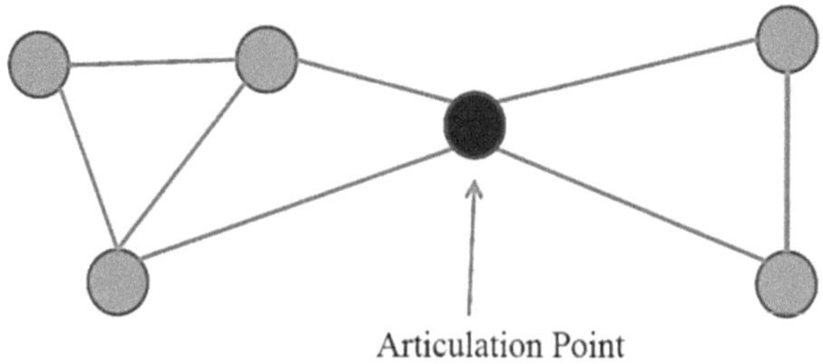

FIGURE 2.6 Articulation point

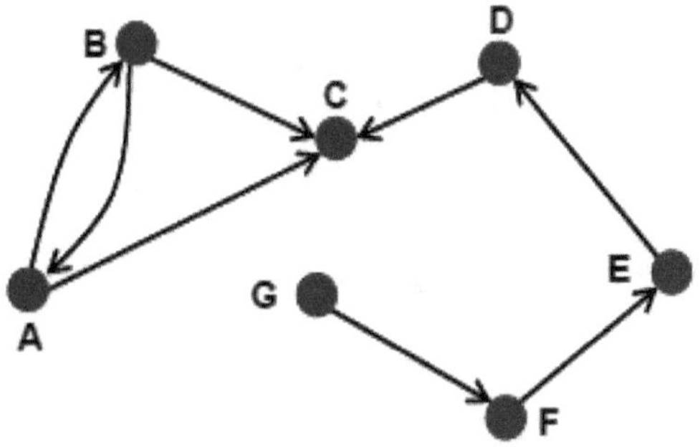

FIGURE 2.7 Directed network

- By linking friends to each other, we obtain the friendship network that plays an important role in the spread of ideas, products and habits and is of major interest to sociology, marketing and health sciences.
- By connecting individuals that have an intimate relationship, we obtain the sexual network of key importance for the spread of sexually transmitted diseases, like AIDS, and of major interest for epidemiology.
- By using phone and email records to connect individuals that call or email each other, we obtain the acquaintance network, capturing a mixture of professional, friendship or intimate links, of importance to communications and marketing.

Connectivity of Directed Graph:

- A **strongly connected directed graph** has a path from each node to every other node and vice versa (e.g., A–B path and B–A path).
- A **weakly connected directed graph** is connected if we disregard the edge directions.

Figure 2.8 is connected but not strongly connected because there is no way to get from F to G by following the edge directions. Given node *v, **what can v reach?*** What

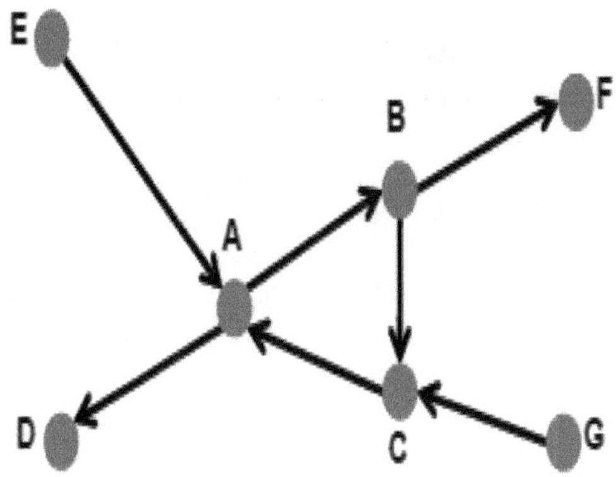

FIGURE 2.8 Connectivity in a directed graph

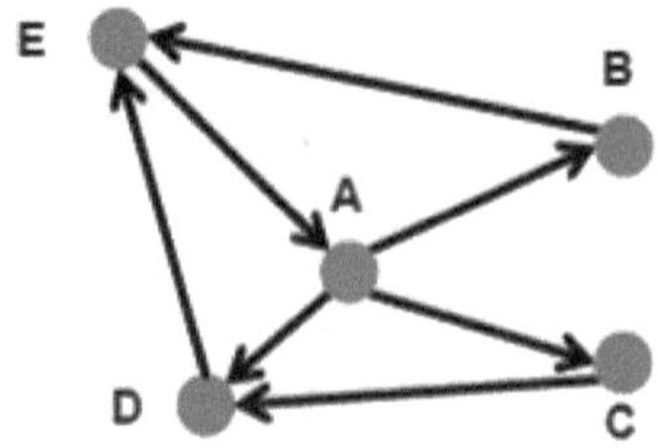

FIGURE 2.9 Strongly connected component

other nodes can reach *v?* To answer these questions, we have to look at in-links and out-links of the node.

$$\text{In}(v) = \{w \mid w \text{ can reach } v\} \tag{2.1}$$

$$\text{Out}(v) = \{w \mid v \text{ can reach } w\} \tag{2.2}$$

For example:

$$\text{In}(A) = \{A, B, C, E, G\}$$

$$\text{Out}(A) = \{A, B, C, D, F\}$$

Directed Graphs

Any directed graph can be expressed in terms of two types:

- **Strongly connected:** Any node can reach any node via a directed path.
- **DAG (directed acyclic graph):** DAG has no cycles: if u can reach v, then v cannot reach u.

$$\text{In}(A) = \text{Out}(A) = \{A, B, C, D, E\}$$

Strongly Connected Component

A strongly connected component (SCC) is a set of nodes *S so that every pair of nodes in S can reach each other.* There is no larger set containing *S with this property.*
SCCs of the graph in Figure 2.10 are: {A, B, C, G}, {D}, {E}, {F}

Fact: Every directed graph is a DAG on its SCCs

- SCCs partition the nodes of G. That is, each node is in exactly one SCC.
- If we build a graph G' whose nodes are SCCs, and with an edge between nodes of G' if there is an edge between corresponding SCCs in G, then G' is a DAG.

We can find an SCC of a graph by:

$$\text{Out}(A) \cap \text{In}(A) \tag{2.3}$$

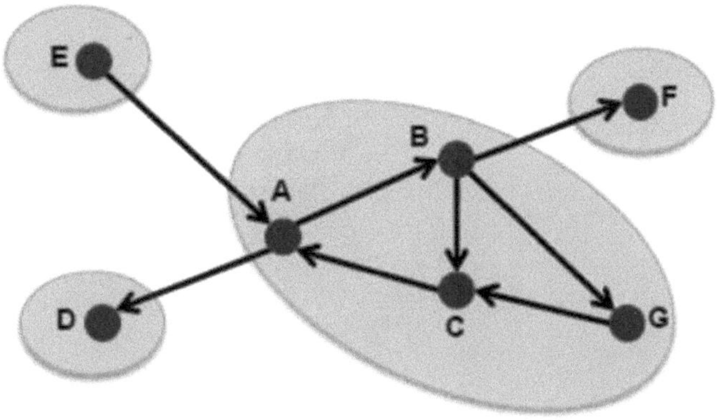

FIGURE 2.10 Example of an SCC

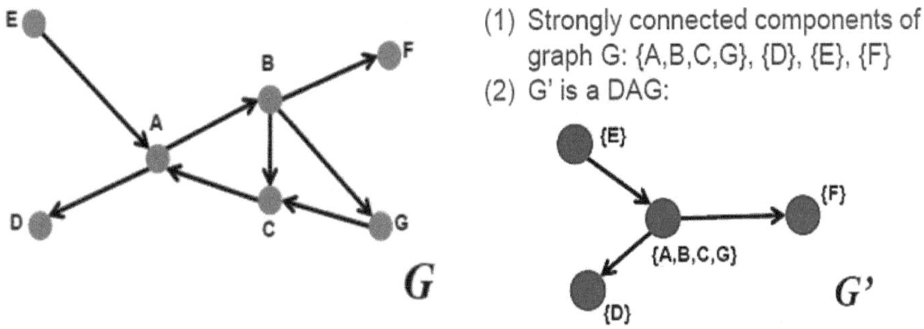

FIGURE 2.11 Example DAG from SCC

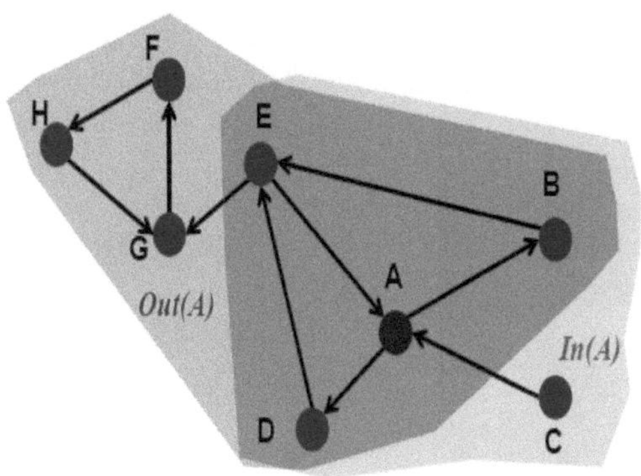

FIGURE 2.12 Example of an SCC

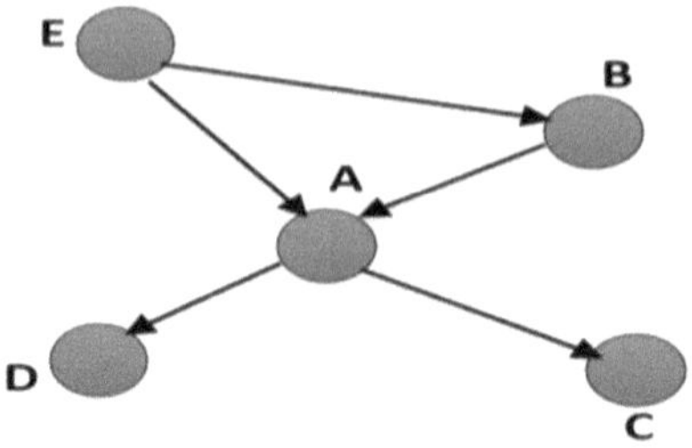

FIGURE 2.13 DAG

Example of an SCC:

$$\text{Out}(A) = \{A, B, D, E, F, G, H\}$$

$$\text{In}(A) = \{A, B, C, D, E\}$$

$$\text{So, SCC}(A) = \text{Out}(A) \cap \text{In}(A) = \{A, B, D, E\}$$

2.2.3 Self-Loops

In many networks, nodes do not interact with themselves, so the diagonal elements of the adjacency matrix are zero, $A_{ii} = 0, i = 1,\ldots, N$. In some systems self-interactions are allowed; in such networks, self-loops represent the fact that node i interacts with itself. Examples: WWW, protein interactions.

2.2.4 Multigraph/Simple Graphs

In a multigraph, nodes are permitted to have multiple links (or parallel links) between them. Hence A_{ii} can be any positive integer. Networks that do not allow multiple links are called simple. Multigraph examples: social networks, where we distinguish friendship, family and professional ties.

2.2.5 Weighted Network

A network whose links have a defined weight, strength or flow parameter. The elements of the adjacency matrix are $A_{ij} = w_{ij}$ if there is a link with weight w_{ij} between them. Examples: collaboration, internet roads

For unweighted (binary) networks, the adjacency matrix only indicates the presence ($A_{ij} = 1$) or the absence $\left(A_{ij} = 0\right)$ of a link. Examples: mobile phone calls, email network, friendship, hyperlink.

2.2.6 Complete Graph (Clique)

In a complete graph, or a clique, all nodes are connected to each other. Examples: actors in the cast of the same movie, because they are all linked to each other in the actor network.

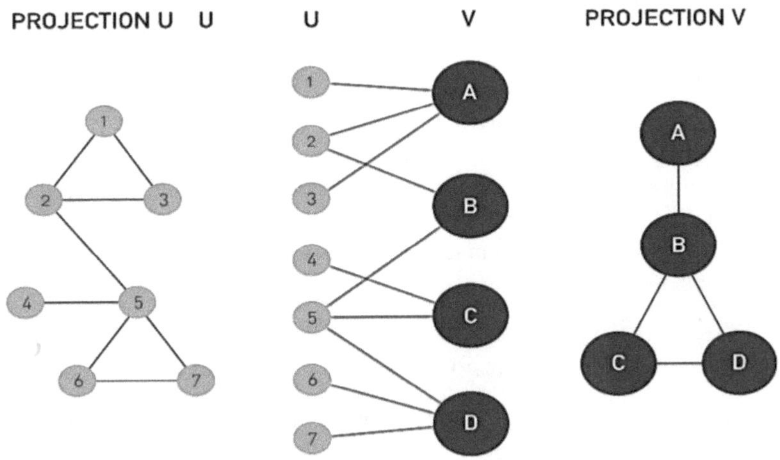

FIGURE 2.14 Bipartite network [1]

2.2.7 Bipartite Graph

A bipartite graph (or bigraph) is a network whose nodes can be divided into two disjoint sets U and V such that each link connects a U-node to a V-node. In other words, if we colour the U-nodes green and the V-nodes purple, then each link must connect nodes of different colours (Figure 2.11). We can generate two projections for each bipartite network. The first projection connects two U-nodes by a link if they are linked to the same V-node in the bipartite representation. The second projection connects the V-nodes by a link if they connect to the same U-node (Figure 2.14).

A bipartite network has two sets of nodes, U and V. Nodes in the U-set connect directly only to nodes in the V-set. Hence there are no direct U–U or V–V links. The figure shows the two projections that we can generate from any bipartite network. Projection U is obtained by connecting two U-nodes to each other if they link to the same V-node in the bipartite representation. Projection *V* is obtained by connecting two V-nodes to each other if they link to the same U-node in the bipartite network.

In network theory, we encounter numerous bipartite networks. A well-known example is the Hollywood actor network, in which one set of nodes corresponds to movies (U) and the other to actors (V). A movie is connected to an actor if the actor features in that movie. One projection of this bipartite network is the actor network, in which two nodes are connected to each other if they played in the same movie. The other projection is the movie network, in which two movies are connected if they share at least one actor in their cast [2, 3].

In network science, we often distinguish networks by some elementary property of the underlying graph. Here, we summarize the most commonly encountered network types. We also list real systems that share the particular property. Note that many real networks combine several of these elementary network characteristics. For example, the WWW is a directed multigraph with self-interactions; the mobile call network is directed and weighted, without self-loops.

2.3 REPRESENTATION OF THE NETWORKS

2.3.1 Adjacency Matrix

A complete description of a network requires us to keep track of its links. The simplest way to achieve this is to provide a complete list of the links. For example, the network of Figure 2.2 is uniquely described by listing its four links: {(1, 2), (1, 3), (2, 3), (2, 4)}. For mathematical purposes, we often represent a network through its adjacency matrix. The adjacency matrix of a directed network of N nodes has N rows and N columns, its elements being:

$A_{ij} = 1$ if there is a link pointing from node j to node i
$A_{ij} = 0$ if nodes i and j are not connected to each other.

The adjacency matrix of an undirected network has two entries for each link, for example, link (1, 2) is represented as $A_{12} = 1$ and $A_{21} = 1$. Hence, the adjacency matrix of an undirected network is symmetric, $A_{ij} = A_{ji}$

The degree k_i of node i can be directly obtained from the elements of the adjacency matrix. For undirected networks, a node's degree is a sum over either the rows or the columns of the matrix:

Adjacency Matrix:

$$A_{ij} = \begin{matrix} A_{11} & A_{12} & A_{13} & A_{14} \\ A_{21} & A_{22} & A_{23} & A_{24} \\ A_{31} & A_{32} & A_{33} & A_{34} \\ A_{41} & A_{42} & A_{43} & A_{44} \end{matrix}$$

Undirected Network Representation:

$$A_{ij} = \begin{matrix} 0 & \mathbf{1} & 1 & 0 \\ \mathbf{1} & \mathbf{0} & \mathbf{1} & \mathbf{1} \\ 1 & \mathbf{1} & 0 & 0 \\ 0 & \mathbf{1} & 0 & 0 \end{matrix}$$

$$\text{Degree of node } 2 = k_2 = \sum_{j=1}^{4} A_{2j} = \sum_{i=1}^{4} A_{i2} = 3$$

$$A_{ij} = A_{ji};\ A_{ii} = 0$$

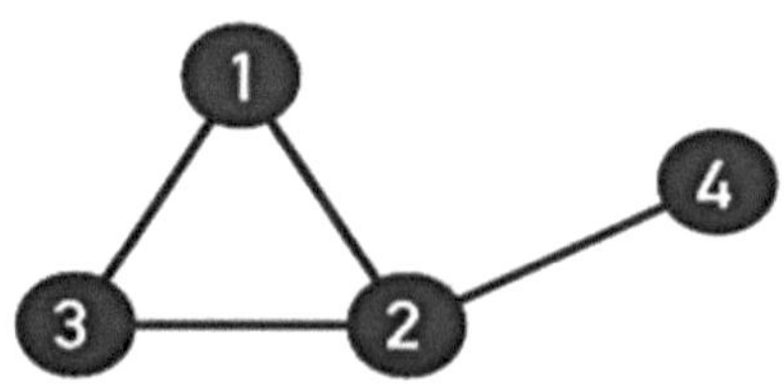

FIGURE 2.15 Undirected network

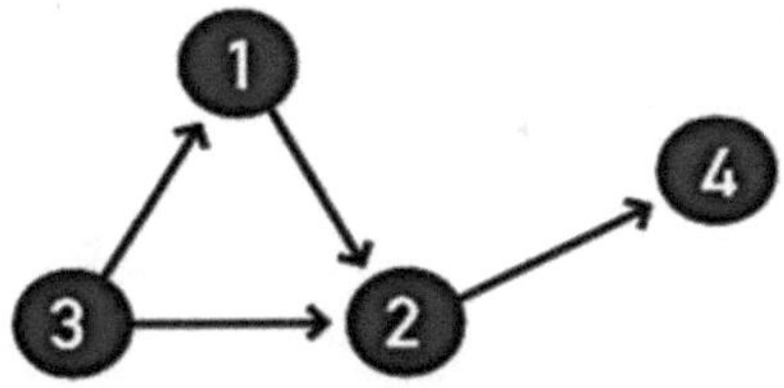

FIGURE 2.16 Directed network

Number of links in an undirected network:

$$L = \frac{1}{2}\sum_{i,j=1}^{N} A_{ij} \tag{2.4}$$

Average degree in an undirected network:

$$\langle k \rangle = \frac{2L}{N} \tag{2.5}$$

Directed Network Representation:

$$A_{ij} = \begin{matrix} 0 & \mathbf{0} & 1 & 0 \\ \mathbf{1} & \mathbf{0} & \mathbf{1} & \mathbf{0} \\ 0 & \mathbf{0} & 0 & 0 \\ 0 & \mathbf{1} & 0 & 0 \end{matrix}$$

$$\text{Degree of node } 2: k_2^{\text{in}} = \sum_{j=1}^{4} A_{2j} = 2;\ k_2^{\text{out}} = \sum_{i=1}^{4} A_{i2} = 1$$

$$A_{ij} \neq A_{ji};\ A_{ii} = 0$$

Number of links in a directed network:

$$L = \sum_{i,j=1}^{N} A_{ij} \tag{2.6}$$

Average degree in a directed network:

$$\langle k^{\text{in}} \rangle = \langle k^{\text{out}} \rangle = \frac{L}{N} \tag{2.7}$$

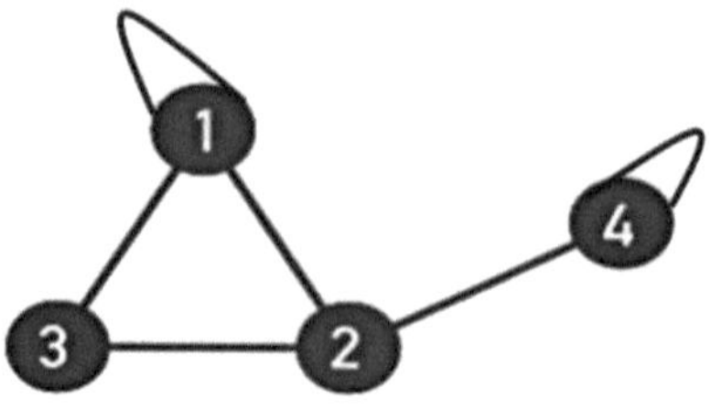

FIGURE 2.17 Self-loop network

Self-loop Network Representation:

$$A_{ij} = \begin{matrix} 1 & 1 & 1 & 0 \\ 1 & 0 & 1 & 1 \\ 1 & 1 & 0 & 0 \\ 0 & 1 & 0 & 1 \end{matrix}$$

$$A_{ij} = A_{ji};\ \ni i,\ A_{ii} \neq 0$$

Number of links in self-loop network:

$$L = \frac{1}{2} \sum_{i,j=1,i \neq j}^{N} A_{ij} + \sum_{i=1}^{N} A_{ii} \tag{2.8}$$

Multigraph Network Representation:

$$A_{ij} = \begin{matrix} 0 & 2 & 1 & 0 \\ 2 & 0 & 1 & 3 \\ 1 & 1 & 0 & 0 \\ 0 & 3 & 0 & 0 \end{matrix}$$

$$A_{ij} = A_{ji};\ A_{ii} = 0$$

Number of links in a multigraph network:

$$L = \frac{1}{2} \sum_{i,j=1}^{N} A_{ij} \tag{2.9}$$

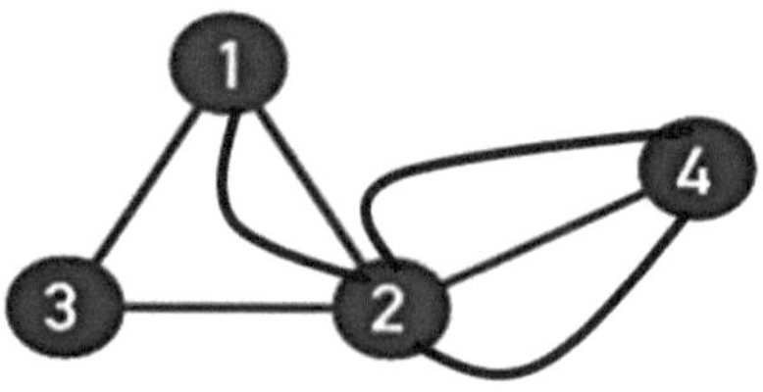

FIGURE 2.18 Multigraph (undirected)

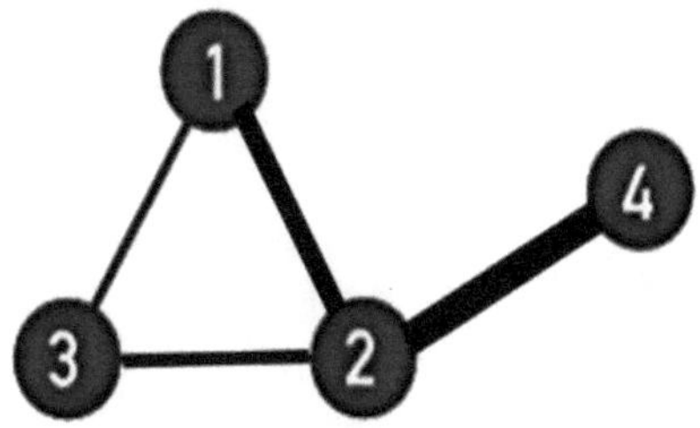

FIGURE 2.19 Weighted network

Average degree in a multigraph network:

$$\langle k \rangle = \frac{2L}{N} \tag{2.10}$$

Weighted Network Representation:

$$A_{ij} = \begin{matrix} 0 & 2 & 0.5 & 0 \\ 2 & 0 & 1 & 3 \\ 0.5 & 1 & 0 & 0 \\ 0 & 4 & 0 & 0 \end{matrix}$$

$$A_{ij} = A_{ji};\ A_{ii} = 0$$

Number of links in a weighted network:

$$L = \frac{1}{2}\sum_{i,j=1}^{N} A_{ij} \tag{2.11}$$

Average degree in a weighted network [1, 4]:

$$\langle k \rangle = \frac{2L}{N} \tag{2.12}$$

Complete (Undirected) Network Representation:

$$A_{ij} = \begin{matrix} 0 & 1 & 1 & 1 \\ 1 & 0 & 1 & 1 \\ 1 & 1 & 0 & 1 \\ 1 & 1 & 1 & 0 \end{matrix}$$

$$A_{i \neq j} = 1;\ A_{ii} = 0$$

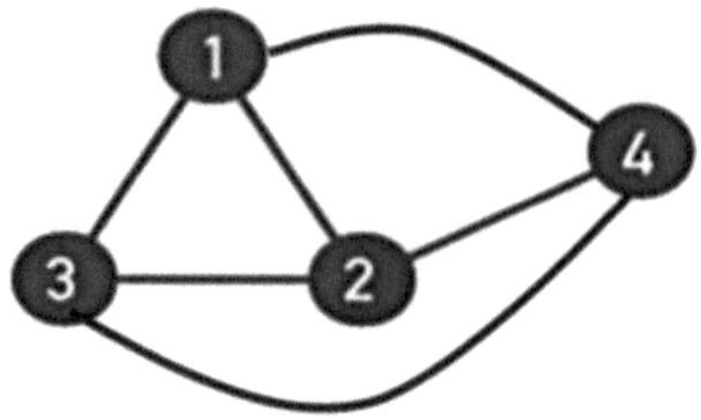

FIGURE 2.20 Complete network

Number of links in a complete network:

$$L = L_{max} = \frac{N(N-1)}{2} \tag{2.13}$$

Average degree in a complete network:

$$\langle k \rangle = N - 1 \tag{2.14}$$

2.3.2 Real Networks are Sparse

In real networks, the number of nodes (N) and links (L) can vary widely. For example, the neural network of the worm *C. elegans*, the only fully mapped nervous system of a living organism, has $N = 302$ neurons (nodes). In contrast, the human brain is estimated to have about a hundred billion ($N \approx 1011$) neurons. The genetic network of a human cell has about 20,000 genes as nodes; the social network consists of 7 billion individuals ($N \approx 7 \times 109$) and the WWW is estimated to have over 1 trillion web documents ($N > 1012$).

These wide differences in size are noticeable in Table 2.1, which lists N and L for several network maps. Some of these maps offer a complete wiring diagram of the system that they describe (like the actor network or the *E. coli* metabolism), while others are only samples, representing a subset of the full network (like the *WWW* or the mobile call graph).

Table 2.1 indicates that the number of links also varies widely. In a network of N nodes, the number of links can change between $L = 0$ and

$$L_{max} = \frac{N}{2} = \frac{N(N-1)}{2} \tag{2.15}$$

where L_{max} is the total number of links present in a complete graph of size N (Figure 2.21). In a complete graph, each node is connected to every other node.

2.3.3 Complete Graph

A complete graph with $N = 16$ nodes and $L_{max} = 120$ links, as predicted by equation. The adjacency matrix of a complete graph is $A_{ij} = 1$ for all i, $j = 1,2,\ldots N$ and $A_{ii} = 0$. The average degree of a complete graph is $\langle k \rangle = N - 1$. A complete graph is often called a clique, a term frequently used in community identification.

TABLE 2.1
Various Real-Time Network Representations

Network	Nodes	Links	Directed/ Undirected	N	L	$\langle K \rangle$
Internet	Routers	Internet connections	Undirected	192,244	609,066	6.34
WWW	Web pages	Links	Directed	325,729	1,497,134	4.60
Power Grid	Power plants, transformers	Cables	Undirected	4,941	6,594	2.67
Mobile Phone Calls	Subscribers	Calls	Directed	36,595	91,826	2.51
Email	Email addresses	Emails	Directed	57,194	103,731	1.81
Science Collaboration	Scientists	Co-authorships	Undirected	23,133	93,437	8.08
Actor Network	Actors	Co-acting	Undirected	702,388	29,397,908	83.71
Citation Network	Papers	Citations	Directed	449,673	4,689,479	10.43
E. coli Metabolism	Metabolites	Chemical reactions	Directed	1,039	5,802	5.58
Protein Interactions	Proteins	Binding interactions	Undirected	2,018	2,930	2.90

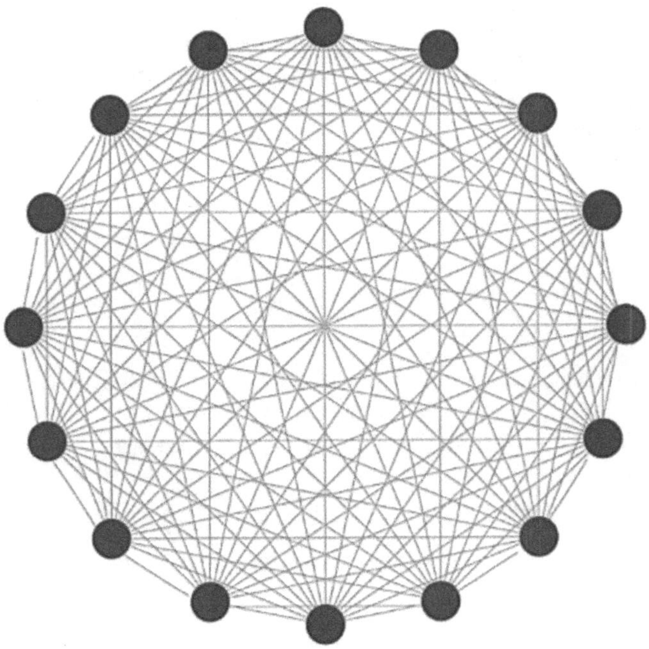

FIGURE 2.21 Complete network [1]

In real networks L is much smaller than L_{max}, reflecting the fact that most real networks are sparse. We call a network sparse if $\ll L_{max}$. For example, the WWW graph in Table 2.1 has about 1.5 million links. Yet, if the WWW were to be a complete graph, it should have $L_{max} \approx 5 \times 1{,}010$ links according to equation. Consequently, the web graph has only a 3×10^{-5} fraction of the links it could have. This is true for all of the networks in Table 2.1. One can check that their number of links is only a tiny fraction of the expected number of links for a complete graph of the same number of nodes.

The sparsity of real networks implies that the adjacency matrices are also sparse. Indeed, a complete network has $A_{ij} = 1$, for all (i, j) (i.e., each of its matrix elements are equal to one). In contrast, in real networks, only a tiny fraction of the matrix elements is non-zero. Sparseness has important consequences on the way that we explore and store real networks. For example, when we store a large network in our computer, it is better to store only the list of links (i.e., elements for which $A_{ij} \neq 0$), rather than the full adjacency matrix, because an overwhelming fraction of the A_{ij} elements are zero. Hence, the matrix representation will block a huge chunk of memory, filled mainly with zeros.

2.4 NETWORK PROPERTIES

In order to understand the characterization of a network, the following properties needs to be understood.

2.4.1 Node Degree

A key property of each node is its degree. It represents the number of links it has to other nodes. Example: the degree can represent the number of mobile phone contacts an individual has in the call graph (i.e., the number of different individuals the person has talked to), or the number of citations that a research paper gets in the citation network.

We denote with k_i the degree of the ith node in the network. For example, for the undirected networks in Figure 2.2, we have $k_1 = 2$, $k_2 = 3$, $k_3 = 2$, $k_4 = 1$. In an undirected network, the total number of links, L, can be expressed as the sum of the node degrees

$$L = \frac{1}{2}\sum_{i=1}^{N} k_i \tag{2.16}$$

Here, the $1/2$ factor corrects for the fact that in the sum each link is counted twice. For example, the link connecting the nodes 2 and 4 in Figure 2.2 will be counted once in the degree of node 1 and once in the degree of node 4.

2.4.2 Average Degree

An important property of a network is its average degree, which for an undirected network is

$$\langle k \rangle = \frac{1}{N}\sum_{i=1}^{N} k_i = \frac{2L}{N} \tag{2.17}$$

In directed networks, we distinguish between incoming degree, k_i^{in}, representing the number of links that point to node i, and outgoing degree, k_i^{out}, representing the number of links that point from node i to other nodes. Finally, a node's total degree, k_i, is given by

$$k_i = k_i^{\text{in}} + k_i^{\text{out}} \tag{2.18}$$

For example, on the WWW, the number of pages that a given document points to represents its outgoing degree, k^{out}, and the number of documents that point to it represents its incoming degree, k^{in}. The total number of links in a directed network is

$$L = \sum_{i=1}^{N} k_i^{\text{in}} = \sum_{i=1}^{N} k_i^{\text{out}} \tag{2.19}$$

The 1/2 factor seen in the equation of an undirected network is now absent, as for directed networks, the two sums separately count the outgoing and the incoming degrees. The average degree of a directed network is

$$\langle k_{\text{in}} \rangle = \frac{1}{N}\sum_{i=1}^{N} k_i^{\text{in}} = \langle k_{\text{out}} \rangle = \frac{1}{N}\sum_{i=1}^{N} k_i^{\text{out}} = \frac{L}{N} \tag{2.20}$$

2.4.3 Degree Distribution

The degree distribution, p_k, provides the probability that a randomly selected node in the network has degree (k). Since p_k is a probability, it must be normalized, that is

$$\sum_{k=1}^{\infty} p_k = 1 \tag{2.21}$$

For a network with N nodes, the degree distribution is the normalized histogram (Figure 2.19) given by

$$p_k = \frac{N_k}{N} \tag{2.22}$$

where N_k is the number of degree(k) nodes. Hence the number of degree(k) nodes can be obtained from the degree distribution as $N_k = Np_k$.

The degree distribution has assumed a central role in network theory following the discovery of scale-free networks. One reason is that the calculation of most network properties requires us to know p_k. For example, the average degree of a network can be written as

$$\langle k \rangle = \sum_{k=0}^{\infty} kp_k \tag{2.23}$$

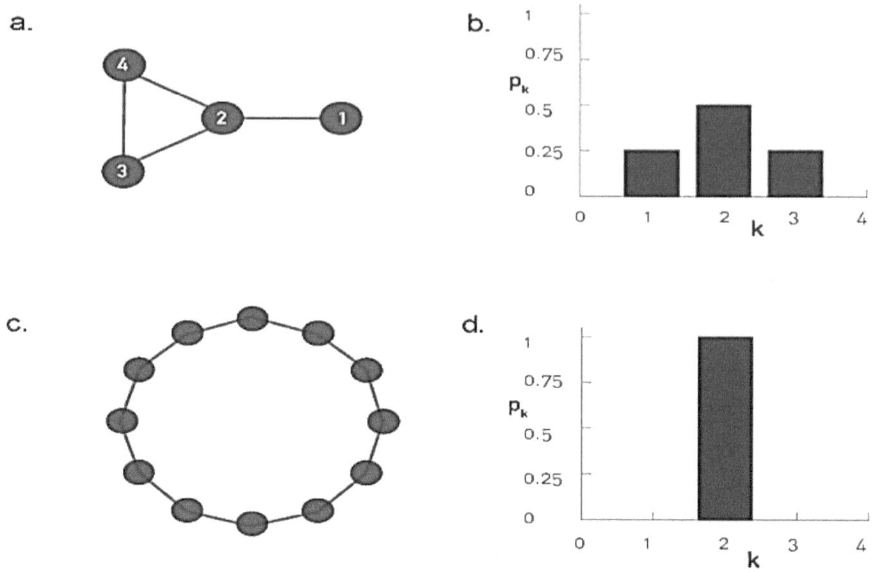

FIGURE 2.22 Degree distribution [1]

The other reason is that the precise functional form of p_k determines many network phenomena, from network robustness to the spread of viruses.

Degree distribution example: the degree distribution of a network is provided by the ratio

$$p_k = \frac{N_k}{N}$$

For the network in Figure 2.22(a) with $N = 4$, the degree distribution is shown in Figure 2.22(b).

We have $p_1 = \frac{1}{4}$ (one of the four nodes has degree $k_1 = 1$), $p_2 = \frac{2}{4} = \frac{1}{2}$ (two nodes have $k_3 = k_4 = 2$) and $p_3 = \frac{1}{4}$ (as $k_2 = 3$). Because we lack nodes with degree $k = 3$, $p_k = 0$ for any $k = 3$.

A one-dimensional lattice for which each node has the same degree $k = 2$.

The degree distribution of Figure 2.22(c) is a Kronecker's delta function, $p_k = \delta(k - 2)$.

2.4.4 Paths and Distance in Graph

Physical distance plays a key role in determining the interactions between the components of physical systems. For example, the distance between two atoms in a crystal or between two galaxies in the universe determine the forces that act between them.

In networks, physical distance is replaced by path length. A path is a route that runs along the links of the network. A path's length represents the number of links

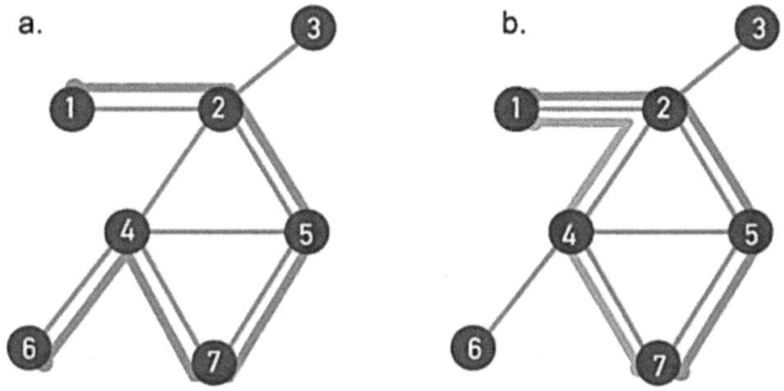

FIGURE 2.23 Path [1]

the path contains (Figure 2.23[a]). Note that some texts require that each node a path visits is distinct. In network science, paths play a central role.

Paths

A path between nodes i_0 and i_n is an ordered list of n links $P = \{(i_0, i_1), (i_1, i_2), (i_2, i_3), \ldots, (i_{n-1} - i_n\}$. The length of this path is n. The path shown in Figure 2.23(a) follows the route $1 \rightarrow 2 \rightarrow 5 \rightarrow 7 \rightarrow 4 \rightarrow 6$, hence its length is $n = 5$.

The shortest paths between nodes 1 and 7, or the distance d_{17}, correspond to the path with the fewest number of links that connect nodes 1 to 7. There can be multiple paths of the same length, as illustrated by the two paths shown in Figure 2.23. The network diameter is the largest distance in the network, being $d_{max} = 3$ here.

Shortest Path (Geodesic Path, *d*)

The path with the shortest distance d between two nodes. We also call d the distance between two nodes. Note that the shortest path does not need to be unique: between nodes 1 and 4, we have two shortest paths, $1 \rightarrow 2 \rightarrow 3 \rightarrow 4$ and $1 \rightarrow 2 \rightarrow 5 \rightarrow 4$ having the same length $d_{1,4} = 3$.

Diameter (d_{max})

The longest shortest path in a graph, or the distance between the two furthest nodes. In the graph shown in Figure 2.24, the diameter is between nodes 1 and 4, hence $d_{max} = 3$.

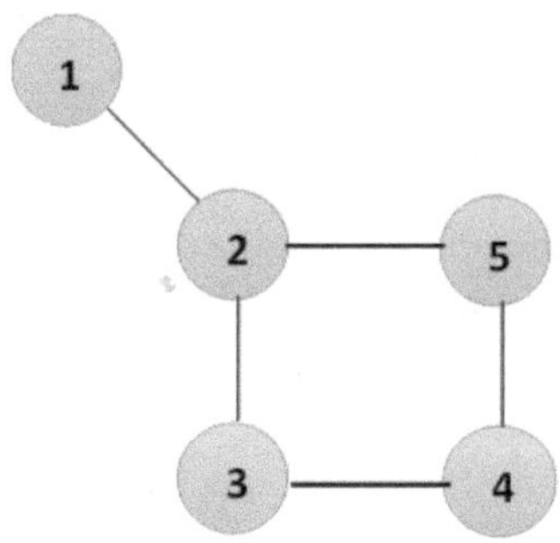

FIGURE 2.24 Example graph for shortest path

Cycle

A path with the same start and end node. In Figure 2.23(a), we only have one cycle.

Eulerian Path

A path that traverses each link exactly once. Figure 2.23(b) shows two such Eulerian paths

Hamiltonian Path

A path that visits each node exactly once. We show two Hamiltonian paths in Figure 2.23(b).

2.4.5 Shortest Path

The shortest path between nodes i and j is the path with the fewest number of links (Figure 2.23 [b]). The shortest path is often called the distance between nodes i and j, and is denoted by d_{ij}, or simply d. We can have multiple shortest paths of the same length d between a pair of nodes (Figure 2.23[b]). The shortest path never contains loops or intersects itself.

In an undirected network $d_{ij} = d_{ji}$ (i.e., the distance between node i and j is the same as the distance between node i and j). In a directed network, often $d_{ij} \neq d_{ji}$. Furthermore, in a directed network, the existence of a path from node i to node j does not guarantee the existence of a path from j to i.

In real networks, we often need to determine the distance between two nodes. For a small network, like the one shown in Figure 2.24, this is an easy task. For a network with millions of nodes, finding the shortest path between two nodes can be rather time-consuming. The length of the shortest path and the number of such paths can be formally obtained from the adjacency matrix. In practice, we use the breadth first search (BFS) algorithm [1, 5, 6].

2.4.6 Network Diameter

The diameter of a network, denoted by d_{max}, is the maximum shortest path in the network. In other words, it is the largest distance recorded between any pair of nodes. One can verify that the diameter of the network shown in Figure 2.24 is $d_{max} = 3$. For larger networks, the diameter can be determined using the BFS algorithm.

2.4.7 Average Path Length

The average path length, denoted by ⟨d⟩, is the average distance between all pairs of nodes in the network. For a directed network of N nodes, ⟨d⟩ is

$$d = \frac{1}{N(N-1)} \sum_{i,j=1,N;i \neq j} d_{i,j} \tag{2.24}$$

Here, $d_{i,j}$ represents the length of shortest path exists between two vertices. So, we take sum of all shortest paths between all vertices and divide number of all possible paths.

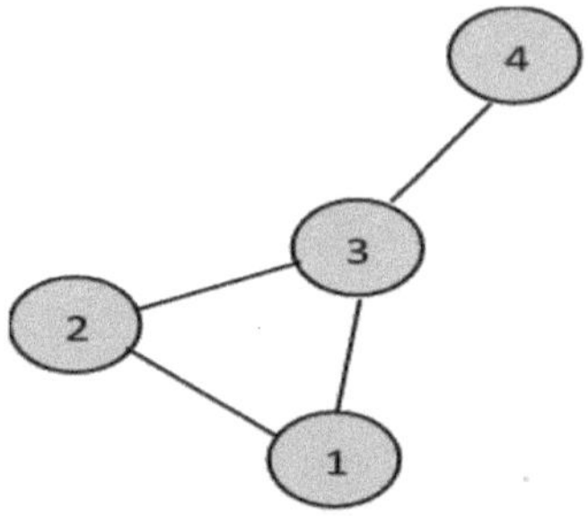

FIGURE 2.25 Average path length example

Example:

Consider the graph shown in Figure 2.25. Now we calculate shortest path for each pair of vertices:

d (1,2) = 1	d (1,3) = 1	d (1,4) = 2
d (2,1) = 1	d (2,3) = 1	d (2,4) = 2
d (3,1) = 1	d (3,2) = 1	d (3,4) = 1
d (4,1) = 2	d (4,2) = 2	d (4,3) = 1

By putting in a formula, we get:

$$\frac{1}{4(4-1)}(1+1+2+1+1+2+1+1+1+2+2+1) = \frac{1}{12}(16) = \frac{4}{3} = 1.33$$

2.4.8 Clustering Coefficient

The clustering coefficient captures the degree to which the neighbours of a given node link to each other. For a node i with degree k_i the local clustering coefficient is defined as [1]

$$C_i = \frac{2L_i}{k_i(k_i-1)} \tag{2.25}$$

where L_i represents the number of links between the k_i neighbours of node i. Note that C_i is between 0 and 1 (Figure 2.26[a]):

$C_i = 0$ if none of the neighbours of node i link to each other.
$C_i = 1$ if the neighbours of node i form a complete graph (i.e., they all link to each other).

C_i is the probability that two neighbours of a node link to each other. Consequently $C = 0.5$ implies that there is a 50% chance that two neighbours of a node are linked.

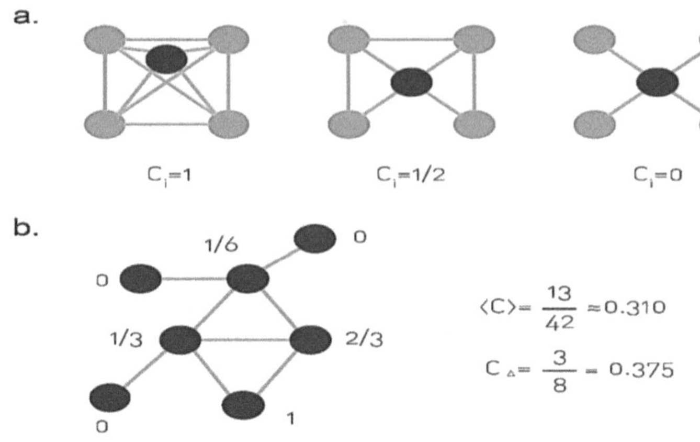

FIGURE 2.26 Clustering coefficient example [1]

In summary, C_i measures the network's local link density: The more densely interconnected the neighbourhood of node i, the higher is its local clustering coefficient [1, 7].

The local clustering coefficient, C_i, of the central node with degree $k_i = 4$ for three different configurations of its neighbourhood. The local clustering coefficient measures the local density of links in a node's vicinity. Note that for nodes with degrees $k_i = 0,1$; the clustering coefficient is zero.

The degree of clustering of a whole network is captured by the **average clustering coefficient**, $\langle C \rangle$, representing the average of C_i over all nodes $i = 1, \ldots, N$

$$\langle \mathrm{C} \rangle = \frac{1}{N}\sum_{i=1}^{N} C_i \tag{2.26}$$

In line with the probabilistic interpretation ⟨C⟩ is the probability that two neighbours of a randomly selected node link to each other.

The degree of a network's **global clustering** can be also captured by the global clustering coefficient, defined as

$$C_\Delta = \frac{3\times \text{ Number of Triangles}}{\text{Number of connected Triples}} \tag{2.27}$$

where a connected triplet is an ordered set of three nodes ABC such that A connects to B and B connects to C. For example, an A, B, C triangle is made of three triplets, ABC, BCA and CAB. In contrast a chain of connected nodes A, B, C, in which B connects to A and C, but A does not link to C, forms a single open triplet ABC. The factor three in the numerator is due to the fact that each triangle is counted three times in the triplet count. C_Δ is often called the ratio of transitive triplets.

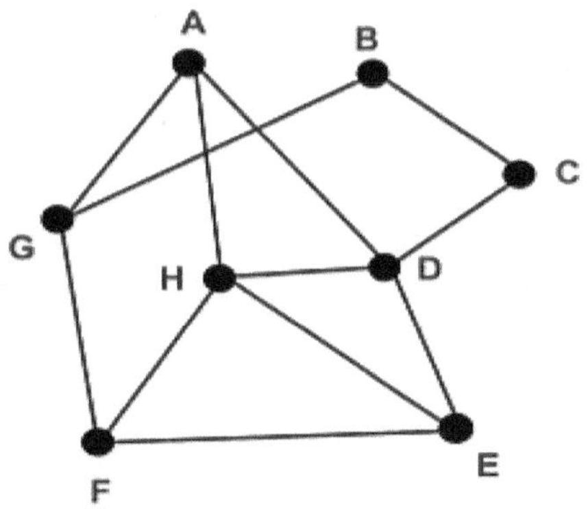

FIGURE 2.27 Graph G for Q1

2.5 PROBLEMS FOR SELF-ASSESSMENT

Q1. Consider graph G in Figure 2.27 and answer the following questions:

a. What is the adjacency matrix of graph G?
b. What is the density of graph G?
c. What is the diameter of graph G?
d. What is the average path length of graph G?
e. What is the node degree of each node of graph G?
f. What is the average degree of the graph of graph G?
g. What is the degree distribution of graph G?
h. What is the clustering coefficient of graph G?
i. What is the maximum value a diameter can have for a connected graph with n vertices?
j. What is the strongly connected component of graph G?

Q2. I have four friends. If my clustering coefficient is 0.33, what is the number of friendships between my friends?

Q3. Consider Figure 2.28, where each edge is labelled as S (strong tie) or W (weak tie) – except the edge connecting B and C. According to the theory of strong and weak ties, with the strong triadic closure assumption, what would be the label for the edge BC?

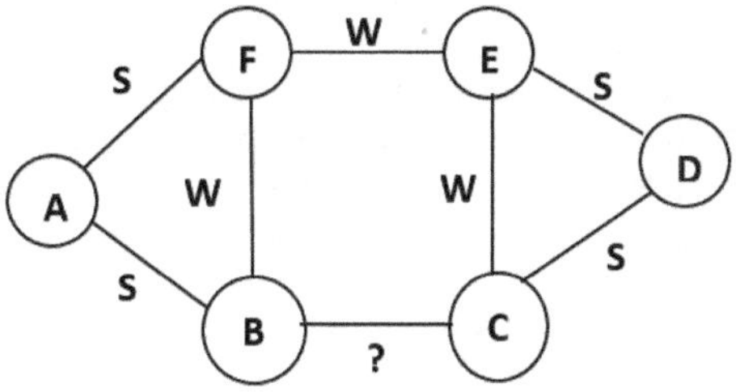

FIGURE 2.28 Graph for Q3

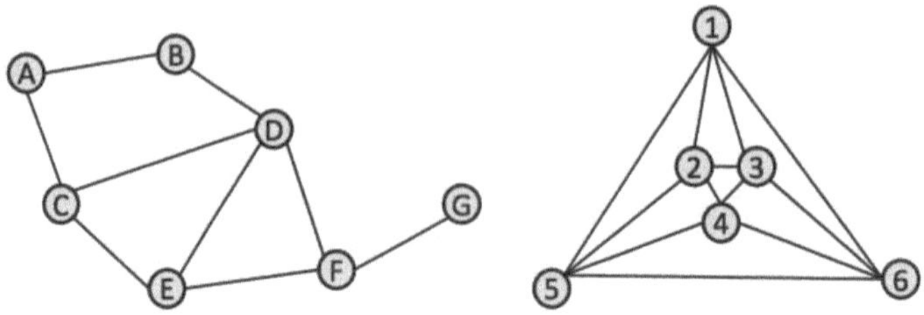

FIGURE 2.29 Graphs for Q4

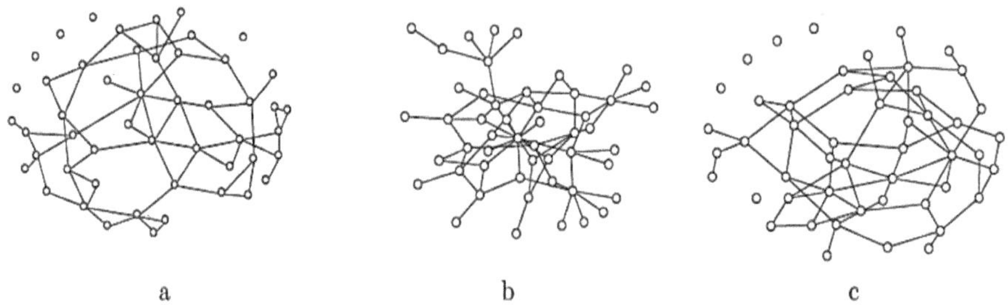

FIGURE 2.30 Graphs for Q5

Q4. In Figure 2.29, what is the diameter of the left network? The right network?

Q5. In Figure 2.30, which of the following graphs has the highest clustering coefficient? (You don't need to actually compute it!) Justify your answer.

REFERENCES

1. Chapter 2: Graph theory. *Network Science Book.* http://networksciencebook.com/chapter/2
2. Barabási, A. L. (2013). Network science. *Philosophical Transactions of the Royal Society A: Mathematical, Physical and Engineering Sciences*, 371(1987), 20120375.
3. Barabási A.-L., & Albert, R. (1999). Emergence of scaling in random networks. *Science*, 286, 509–512.
4. Zhang, B. & Horvath, S. (2005). A general framework for weighted gene coexpression network analysis. *Statistical Applications in Genetics and Molecular Biology*, 4, 17. doi: 10.2202/1544-6115.1128
5. Gilder, G. (1993). Metcalfe's law and legacy. *Forbes ASAP.*
6. Barrat, A., Barthélemy, M., Pastor-Satorras, R., & Vespignani, A. (2004). The architecture of complex weighted networks. *Proceedings of the National Academy of Sciences of the United States of America*, 101, 3747–3752.
7. Onnela, J. P., Saramäki, J. Kertész, J., & Kaski, K. (2005). Intensity and coherence of motifs in weighted complex networks. *Physical Review E*, 71, 065103.

3 Network Models

Network topological analysis network topology is the arrangement of the various elements (links, nodes etc.). Essentially, it is the topological structure of a network. It looks at how to model the topology of large-scale networks, what the organizing principles underlying their topology are and how the topologies of a network affect its robustness against errors and attacks.

Concepts such as degree of distribution, small average distance, large clustering coefficient and strongly connected components have been described in the previous chapter of this book. With this chapter, we aim to discuss various networking models such as the random graph model, small-world model and scale-free model. But before getting into these models and their explanation, let us just have a quick recap of the various features of networks (in context of distribution).

3.1 BASIC FEATURES OF NETWORKS

The understanding of networks requires the conceptualization of various types of distribution, both in a continuous and discrete framework [1].

3.1.1 Continuous Distribution

For a continuous random variable X

Probability density function (PDF) $f\ (x)$:

$$P_r(a \le X \le b) = \int_a^b f(x)dx \tag{3.1}$$

$$f(x) \ge 0, \int_{-\infty}^{\infty} = 1$$

Cumulative distribution function (CDF):

$$F(x) = P_r(X \le x) = \int_{-\infty}^{x} f(y)dy \tag{3.2}$$

3.1.2 DISCRETE DISTRIBUTION

For a discrete random variable X_i

Probability mass function (PMF) $p(x)$:

$$p(x) = P_r(Xi \le x) = \sum_{y \le x} p(y) \tag{3.3}$$

DOI: 10.1201/9781003088066-3

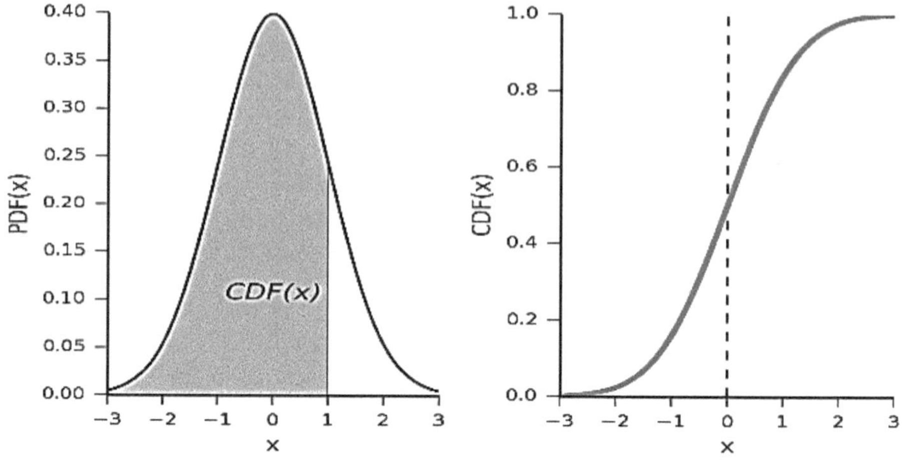

FIGURE 3.1 The shape of a CDF

Once the concept of these distributions becomes clear, one can now understand the power law and Zipf's law [2].

A power law is a 'functional relationship between two quantities, where a relative change in one quantity results in a proportional relative change in the other quantity, independent of the initial size of those quantities: one quantity varies as a power of another'. For instance, considering the area of a square in terms of the length of its side, if the length is doubled, the area is multiplied by a factor of four.[2] A power law can be turned into a linear relationship if we plot the variables on logarithmic axes. Plotting two quantities against each other in this way is how we generally determine if they have a power law relationship.

A power law distribution has the form $Y = k.X^{\alpha}$

where,

X and Y are variables of interest,

α is the law's exponent,

k is a constant.

Any inverse relationship like $Y = X^{-1}$ is also a power law, because a change in one quantity results in a negative change in another.

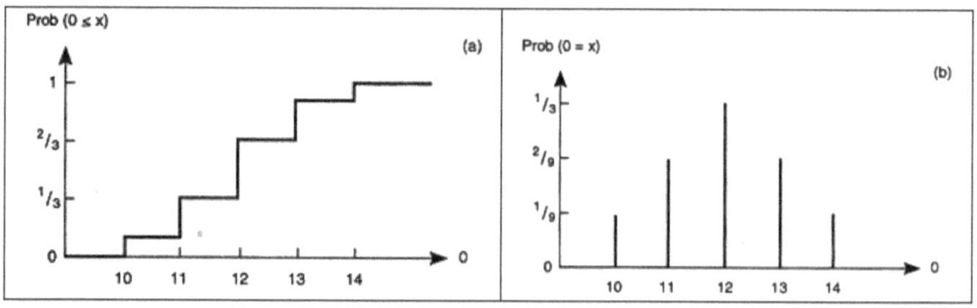

FIGURE 3.2 Curves for a discrete distribution

Let us understand this concept using the following example:

A power law can be used to describe a phenomenon where a small number of items is clustered at the top of a distribution (or at the bottom), taking up 95% of the resources. In other words, it implies a small number of occurrences is common, while larger occurrences are rare. For example, where the distribution of income is concerned, there are very few billionaires; the bulk of the population holds very modest set of income. Other examples include magnitude of earthquakes, size of cities according to population, size of corporations, trading volumes on the stock market, word frequencies and many more where the concentration is very high at one end and very low throughout [3].

Scale invariance, lack of well-defined average value and universality are some of the properties of power law.

Another related law is Zipf's law that usually refers to the frequency of an event relative to its rank. Zipf's law states that, given a list of the most frequent words in an arbitrary book, the most frequent word will appear twice as often as the second most frequent word, which will appear twice as often as the third most frequent etc. In its simplest form, Zipf's law is equal to the power law. Zipf's law gives rise to the Zeta distribution.

3.2 GENERATIVE MODELS

3.2.1 Random Graph Models

Erdös and Rényi proposed the earliest model of a random graph in 1959. The model starts with a NULL graph having n-nodes. Each (n [n – 1])/2 of the pairs of nodes are connected with an edge at random graph with a specified uniform probability p. At a threshold value pc of the probability p, many interesting properties appear in the graph. When $p < pc$, ($pc \sim i/n$), the graph has many disconnected components.

Example: if we have a null graph with $n = 10$ nodes, with a probability $p = 0.2$, an edge is added between any two nodes (e.g., nodes c and j). The process is repeated until a total of ($p.n.$ [n – 1]/2) = 9 edges. This leads to formation of a random graph, which has a small diameter.

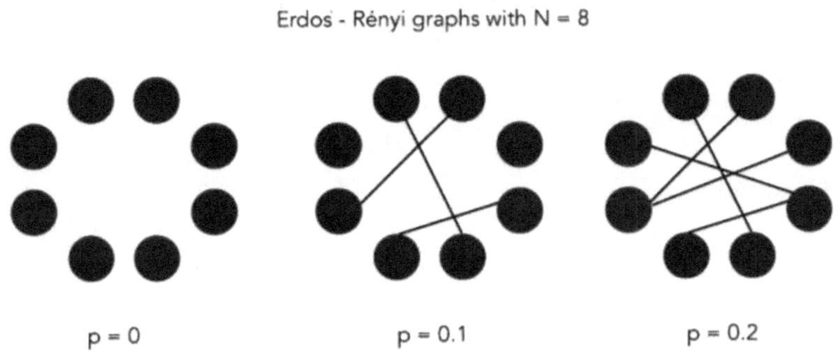

FIGURE 3.3 Erdös–Rényi model [4]

Using NetLogo as modelling tool, this can be understood (described in Chapter 12).

ALGORITHM:

G {E, V}, nodes $n = |V|$, edges $m = |E|$. $G_{n,m}$, a randomly selected graph from the set of $C^m_{n(n-1)/2}$ graphs with n nodes and m edges. $G_{n,p}$, each pair out of $n(n-1)/2$ pairs of nodes are connected with probability p, m – random number.

$$\langle m \rangle = p\frac{n(n-1)}{2} \tag{3.4}$$

$$\langle k \rangle = \frac{1}{n}\sum_i k_i = \frac{2\langle m \rangle}{n} = p(n-1) \approx pn \tag{3.5}$$

$$\rho = \frac{\langle m \rangle}{n(n-1)/2} = p \tag{3.6}$$

Probability that the ith node has a degree 0

$$P(k_i = k) = P(k) = C^k_{n-1} p^k (1-p)^{n-1-k} \tag{3.7}$$

(Bernoulli distribution)

p^k - probability that connects to k nodes (has k-edges)
$(1-p)^{n-1-k}$ - probability that does not connect to any other node
C^k_{n-1} - number of ways to select k nodes out of all to connect to

Statistically, the Bernoulli distribution tends to the Poisson distribution under certain circumstances, for example when $n \to \infty$ at fixed $\langle k \rangle = pn = \lambda$

$$p(k) = \frac{\langle k \rangle^k e^{-\langle k \rangle}}{k!} = \frac{\lambda^k e^{-\lambda}}{k!}$$

$$p(k) = \frac{\lambda^k e^{-\lambda}}{k!} \tag{3.8}$$

$$P(k_i = k) = \frac{\lambda^k e^{-\lambda}}{k!}, \lambda = pn$$

Phase transition:

Consider $G_{n,p}$ as a function of p
$p = 0$, empty graph
$p = 1$, complete (full) graph

There exists critical p_c, structural changes from $p < p_c$ to $p > p_c$
Gigantic connected component appears at $p > p_c$

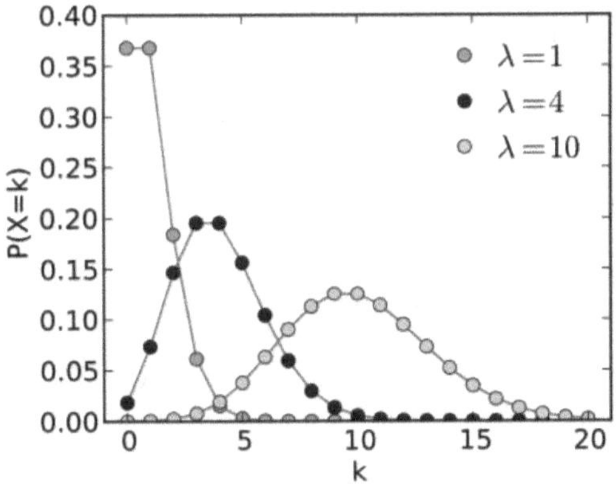

FIGURE 3.4 Poisson distribution [5]

Random graph model

Let u – fraction of nodes that do not belong to GCC. The probability that a node does not belong to GCC

$$u = P(k=1)\cdot u + P(k=2)\cdot u^2 + P(k=3)\cdot u^2 \ldots =$$

$$\sum_{k=0}^{\infty} P(k)u^k = \sum_{k=0}^{\infty} \frac{\lambda^k e^{-\lambda}}{k!} u^k = e^{-\lambda} e^{\lambda u} = e^{\lambda(u-1)} \tag{3.9}$$

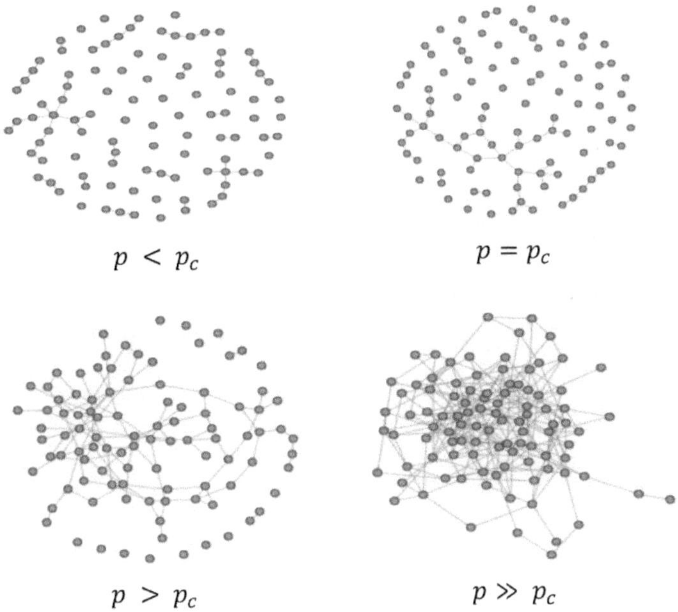

FIGURE 3.5 Random graph model [5]

Let s – fraction of nodes belonging to GCC (size of GCC)

$$s = 1 - u \tag{3.10}$$

$$1 - s = e^{-\lambda s}$$

When $\lambda \to \infty$, $s \to 1$
When $\lambda \to 0$, $s \to 0$

$$(\lambda = pn)_{\text{Le}}$$

$$s = 1 - e^{-\lambda s} \tag{3.11}$$

a non-zero solution exists when $(at\ s = 0)$:

$$\lambda e^{-\lambda s} > 1$$

Critical value:

$$\lambda_c = 1$$

$$\lambda_c = p_c n = 1,\ p_c = \frac{1}{n}$$

Graph $G(n,\ p)$, for $n \to \infty$, critical value $p_c = \frac{1}{n}$

- when $p < p_c$, $(\langle k \rangle < 1)$ there are no components with more than $O(ln\ n)$ nodes, largest component is a tree
- when $p = p_c$, $(\langle k \rangle = 1)$ the largest component has O $(n^{2/3})$ nodes.
- when $p > p_c$, $(\langle k \rangle > 1)$ gigantic component has all $O(n)$ nodes

Critical value: $\langle k \rangle = p_c n = 1$ – on average one neighbour for a node.

Threshold Probabilities

For graph G $(n;\ p)$, threshold probabilities when different subgraphs of g – nodes appear in a random graph.

- $p_c \sim n^{-g/(g-1)}$, having a tree of order g
- $p_c \sim n^{-1}$, having a cycle of order g
- $p_c \sim n^{-2/(g-1)}$, complete subgraph of order g

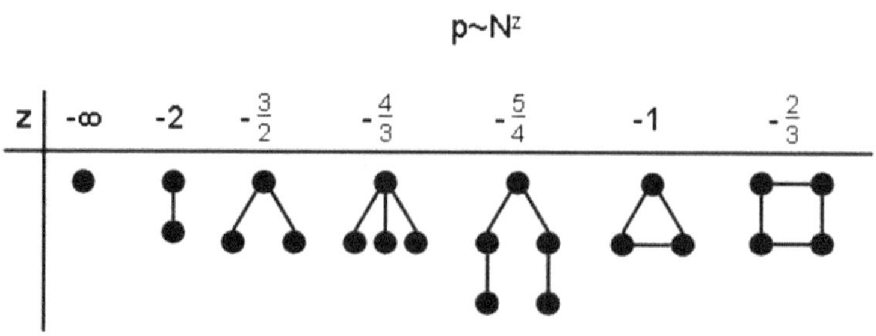

FIGURE 3.6 Emergence of subgraphs

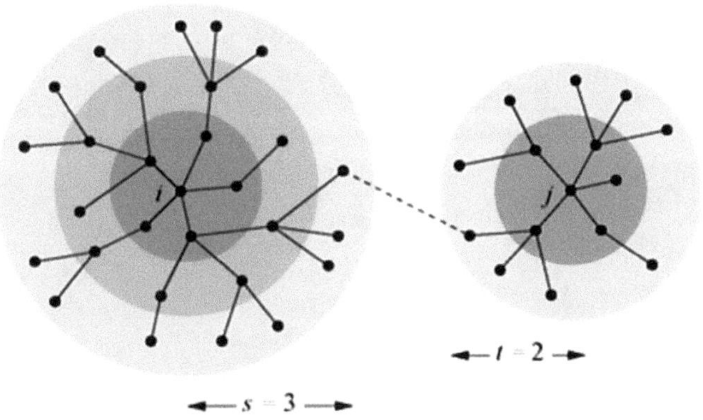

FIGURE 3.7 Graph diameter [5]

Graph Diameter

On average, the number of nodes s steps away from a node $\langle k\rangle^s = \lambda^s$.

If the graph is a tree (GCC, around p_c), $\lambda^d \sim n$, $d \sim \frac{\ln n}{\ln\lambda}$

$P(d_{ij} > s + t + 1)$ – probability, that there is no edge between the surfaces

$P(d_{ij} > s + t + 1) = (1-p)^{\lambda^{s+t}}$, where $\lambda^s\lambda^t$ is total number of possible pairs from different groups

define $l = s + t + 1$

$$P(d_{ij} > l) = (1-p)^{\lambda^{l-1}} = \left(1-\frac{\lambda}{n}\right)^{\lambda^{l-1}}$$

$$\ln P(d_{ij} > l) = \lambda^{l-1}\ln\left(1-\frac{\lambda}{n}\right) = -\frac{\lambda^l}{n}$$

$$P(d_{ij} > l) = \exp\left(-\frac{\lambda^l}{n}\right)$$

The graph diameter is the smallest value l such that $P(d_{ij} > l) = 0$ (i.e., no matter which pair of nodes we pick), there is zero chance to be separated by greater distance, $\lambda^l = an$, should grow faster than n.

$$d = \min(l) = \frac{\ln a}{\ln\lambda} + \frac{\ln n}{\ln\lambda} = A + \frac{\ln n}{\ln\lambda} \tag{3.12}$$

Graph diameter when $p \geq p_c$, $(\lambda = \langle k\rangle = pn)$:

$$d = \frac{\ln n}{\ln\langle k\rangle} \tag{3.13}$$

Clustering Coefficient

$$C(k)=\frac{\#\text{ of links between }NN}{\#\text{ max number of links }NN}=\frac{pk(k-1)/2}{k(k-1)/2}=p$$

$$C=p=\frac{\langle k\rangle}{n} \tag{3.14}$$

When $n\to\infty,\ C\to 0$ [6]

Configuration Model

Select a sequence of nodes with degrees

$$D=k_1,\ k_2,\ k_3,\ \ldots,\ k_n:\sum_i k_i=2m$$

to follow given distribution $P(k)$. For example: 1 1 1 1 1 2 2 2 3 3 3 …

$$P(k)=\frac{\#(k_i=k)}{2m} \tag{3.15}$$

Randomly select two nodes from the sequence and form an edge between them.

Growing (evolving with time) networks: some examples are citation networks, collaboration networks, web and social networks.

In order to understand how a graph evolves with time, we have a growing random graph.

Stochastic growth model:

Suppose at t = 0, there are m unconnected nodes. Thus, we can visualize growth as:

on every time-step $t=\{1,\ 2,\ 3,\ 4\ldots\}$ add a node with m edges $k_i(t=i)=m$
attachment: form m edges with existing nodes uniformly at random,
Expected $i-$node degree at t $\langle k_i(t)\rangle$:

$$k_i(t)=m+\frac{m}{i+1}+\frac{m}{i+2}+\ldots\frac{m}{t}$$

$$k_i(t)=m\left(1+\log\left(\frac{t}{i}\right)\right) \tag{3.16}$$

$$k_i(t)=m\left(1+\log\left(\frac{t}{i}\right)\right),\ m=20,\ i=10,\ 20,\ 40;\ t\geq i$$

$$k_i(t)=m\left(1+\log\left(\frac{t}{i}\right)\right),\ m=20,\ t=50,\ 100,\ 200;\ i\leq t$$

Fraction of nodes with degrees $ki\ (t) < k$ (CDF):

$$F(k) = P(k_i\ (t) < k) = \frac{t-i}{t} = \frac{t - e^{\frac{m-k}{m}}}{t} = 1 - e^{\frac{m-k}{m}}$$

$$P(k) = \frac{d}{dk} F(k) = \frac{1}{m} e^{\frac{m-k}{m}},\ k \geq m \tag{3.17}$$

Mean Field Approximation

Continuous time approximation:

$$k_i\ (t + \delta t) = k_i\ (t) + \frac{m}{t} \delta t \tag{3.18}$$

Differential Equation

$$\frac{dk_i\,(t)}{dt} = \frac{m}{t}$$

with initial conditions: $k_i\,(t = i) = m$

Solution:

$$k_i\ (t) = m\left(1 + \log\left(\frac{t}{i}\right)\right) \tag{3.19}$$

3.2.2 Preferential Attachment Model

The preferential attachment model, proposed by Albert and Barabási in 1999, starts with a small, null graph having a finite number of nodes (n0). At each successive time-step, a new node with a random, but bounded number of outgoing edges (m0) is added to the network. The probability that an edge is added between a new node and an existing node is *d/Σdi* where *di* is the degree of node *i*, and the denominator sum runs over all existing nodes.

Thus, a newly introduced node is more likely to be adjacent to a node with high degree. This is the preferential attachment rule. The small-world models cannot be applied directly to the web because number of nodes (pages) in the web is variable. These models do not accommodate a birth/death process in which new pages are created. They also do not explain how new links are formed (possibly through editing old pages), and how both links and pages can be deleted. The preferential attachment of newly introduced nodes implies that the nodes that are added at early stages of development are more likely to have a high degree.

It is a dynamical growth model; where at; $t = 0$, n_0 unconnected nodes

For growth, we consider that on every step we add a node with m edges $(m \leq n_0)$, $k_i\,(i) = m$

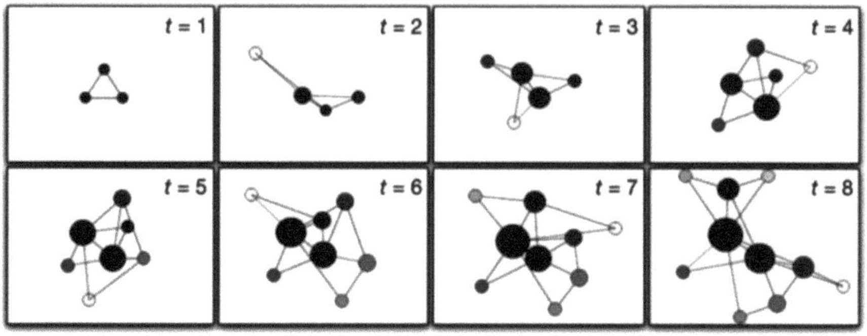

FIGURE 3.8 Preferential attachment models [5]

As per the preferential attachment model; the probability of linking to existing node is proportional to the node degree k_i

$$\Pi k_i = \frac{k_i}{\sum_i k_i} \tag{3.20}$$

After t steps: $n_0 + t$ nodes, mt edges.

For continuous time, mean field approximation, consider the following equation:

$$k_i(t+\delta t) = k_i(t) + m\Pi(k_i)\delta t$$

$$\frac{dk_i(t)}{dt} = m\Pi(k_i) = m\frac{k_i}{\sum_i k_i} = \frac{mk_i}{2mt}$$

$$\frac{dk_i(t)}{dt} = \frac{k_i(t)}{2t}$$

where the initial conditions: $k_i(t) = m$

Solution: $k_i(t) = m\left(\frac{t}{i}\right)^{1/2}$

Time evolution of a node degree

$$k_i(t) = m\left(\frac{t}{i}\right)^{\frac{1}{2}}$$

Nodes with $k_i(t) < k$:

$$m\left(\frac{t}{i}\right)^{\frac{1}{2}} < k$$

$$i > \frac{m^2}{k^2}t$$

Fraction of nodes with $k_i(t) < k$ (CDF)

$$F(k) = P(k_i(t) < k) = \frac{n_0 + t + m^2 t/k^2}{n_0 + t} \approx 1 - \frac{m^2}{k^2}$$

Distribution function:

$$P(k) = \frac{d}{dk} F(k) = \frac{2m^2}{k^3} \tag{3.21}$$

Dynamic Growth

The growth pattern in both the models can be obtained using the following set of equations:

$$BA: k_i(t) = m\left(\frac{t}{i}\right)^{1/2}, \; RG: k_i = m\left(1 + log\left(\frac{t}{i}\right)\right)$$

$$BA: P(k) = \frac{2m^2}{k^3}, \; RG: \frac{1}{m} e^{\frac{m-k}{m}}$$

Power law distribution: $P(k) = \frac{2m^2}{k^3}$

Average path length (analytical result): $\langle L \rangle \sim \log(N) / \log(\log(N))$

Clustering coefficient (numerical result): $C \sim N^{-0.75}$

The third model in this series is a small-world model.

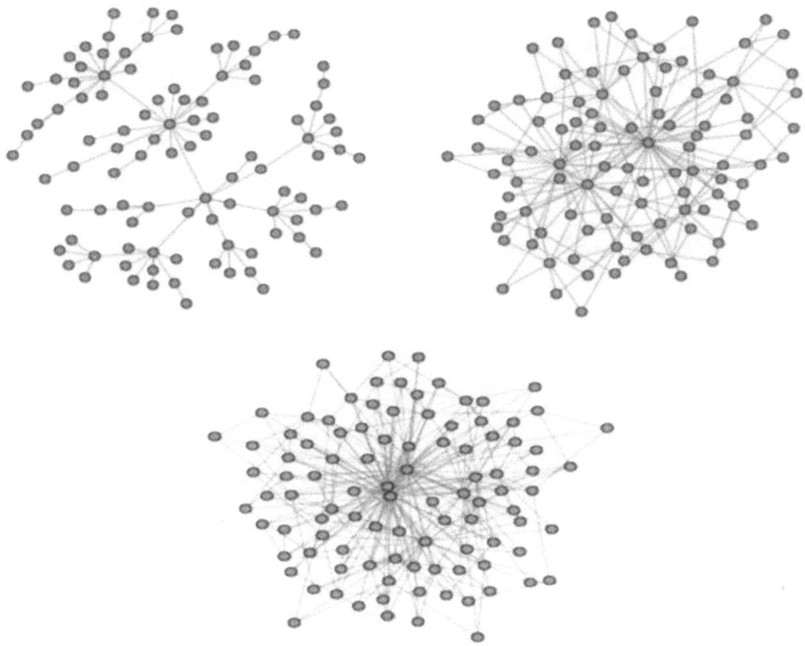

FIGURE 3.9 For m = 1, 2 and 3, respectively

3.2.3 Small-World Model

The small-world model, proposed by Watts and Strogatz in 1998, is a single parameter model, an interpolation between a regular lattice and a random graph. It starts with a regular lattice with N nodes, K nearest neighbours (node degree). It randomly connects with other nodes with *probability* p, forms $pNK = 2$, 'long distance' connections.

When $p = 0$ it is regular lattice. when $p = 1$ it becomes random graph.

This network has two properties:

- Characteristic-path length L that measures the separation between two nodes (global property).
- Clustering coefficient C that measures the cliquishness of the neighbourhood of a node (local property).

Clustering is a measure of the extent to which the neighbourhood of a node forms a complete graph, and it provides the basis for redefining a webpage in a purely WWW structural term.

Watts–Newman–Moore Small-World Network

Analysis of the small-world model has gone well beyond the description provided by the Watts–Strogatz model. The Watts–Newman–Moore model extends the Watts–Strogatz small-world model and is based on undirected graphs. In this model, additional edges are added randomly giving an expected number: $(p - d - n) / 2$ of new edges. Here, p, d and n denote the probability of addition of a new degree of each node in the original ring lattice, and the number of nodes in the ring lattice, respectively. The model exhibits the cliquishness and short paths among nodes, found in social networks. The small-world models cannot be applied directly to the web because the number of nodes (pages) in the web is variable. These models do not accommodate a birth/death process in which new pages are created. They also do not explain how new links are formed (possibly through editing old

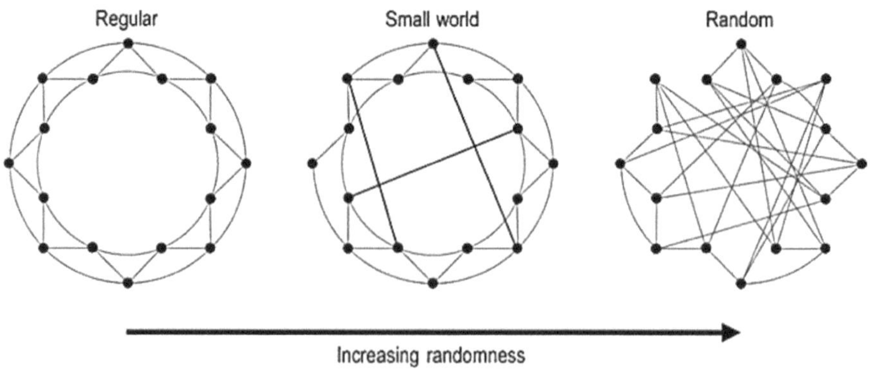

FIGURE 3.10 Regular network, small-world network and random network [7]

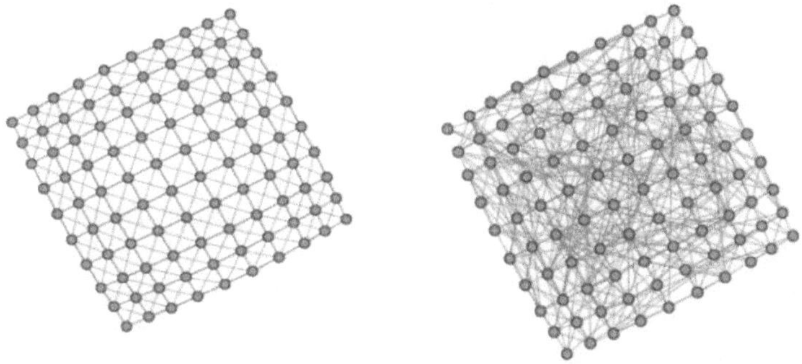

FIGURE 3.11 20% rewiring: avg.path length = 3.58 → avg.path length = 2.32 [5]

pages), and how both links and pages can be deleted. Two new models have been proposed to explain some of the birth/death process: the preferential attachment model proposed by Albert and Barabási and the website growth model proposed by Huberman and Adamic.

Here, the node degree distribution is considered Poisson-like:

- -Ave. path length $\langle L(p) \rangle$:

$p \to 0$, ring lattice, $\langle L(p) \rangle = {}^{2n}\!/_{k}$

$$p \to 1, \text{ random graph, } \langle L(1) \rangle = {}^{\log(n)}\!/_{\log(k)}$$

- Clustering coefficient $C(p)$:

$p \to 0$, ring lattice, $C(0) = \frac{3}{4} = \text{const}$

$$p \to 1, \text{ random graph, } C(1) = {}^{k}\!/_{n}$$

TABLE 3.1 shows comparison of above discussed models.

	Random	**Albert and Barabási model**	**Watts–Strogatz model**	**Empirical networks**
$P(k)$	$\frac{\lambda^k e^{-\lambda}}{k!}$	k^{-3}	Poisson-like	power law
C	$\langle k \rangle / N$	$N^{-0.75}$	const	large
$\langle L \rangle$	$\frac{\log(N)}{\log(k)}$	$\frac{\log(N)}{\log\log(N)}$	$\log N$	small

3.3 SIX DEGREES OF SEPARATION

Six degrees of separation is the theory that any person on the planet can be connected to any other person on the planet through a chain of acquaintances that has no more than five intermediaries. The concept of six degrees of separation is often represented by a graph database, a type of NoSQL database that uses graph theory to store, map and query relationships. Real-world applications of the theory include power grid mapping and analysis, disease transmission mapping and analysis, computer circuitry design and search engine ranking [9]. The six degrees of separation theory was first proposed in 1929 by the Hungarian writer Frigyes Karinthy in a short story called *Chains*. In the 1950s, Ithiel de Sola Pool (MIT) and Manfred Kochen (IBM) set out to mathematically prove the theory. Although they were able to mathematically phrase the question (given a set N of people, what is the probability that each member of N is connected to another member via k_1, k_2, $k_3 \ldots k_n$ links?), after 20 years, they were still unable to solve the problem to their satisfaction [10].

In 1967, American sociologist Stanley Milgram devised a new way to test the theory, which he called 'the small-world problem'. Milgram randomly selected people in the Midwest to send packages to a stranger located in Massachusetts. The senders knew the recipient's name, occupation and general location. Each participant was instructed to send the package to a person who they knew on a first name basis who was most likely, out of all the participant's friends, to know the target personally. That person would do the same, and so on until the package was personally delivered to its target recipient. Although participants expected the chain to include at least a hundred intermediaries, it only took (on average) between five and seven intermediaries for each package to be delivered successfully [7].

Examples of Degree of Separations on Different Platforms

Email Graph

D. Watts (2001), 48,000 senders, $\langle L \rangle \approx 6$

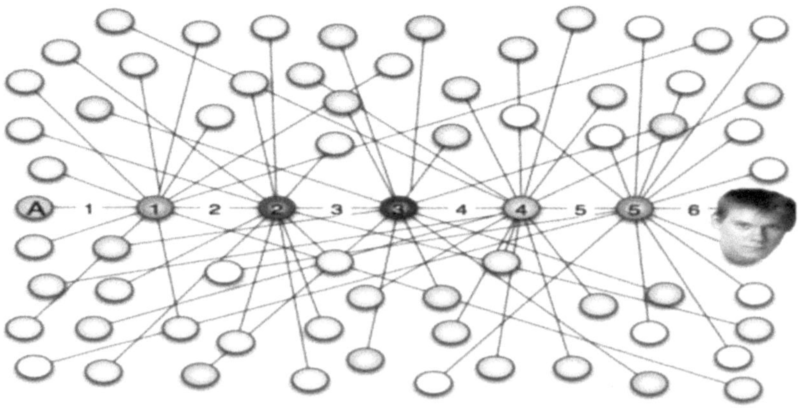

FIGURE 3.12 Six degrees of separation [8]

MSN Messenger Graph

J. Lescovec et al. (2007), 240 million users, $\langle L \rangle \approx 6:6$

Facebook Graph

L. Backstrom et al. (2012), 721 million users, $\langle L \rangle \approx 4:74$

3.4 PROBLEMS FOR SELF-ASSESSMENT

Q1. Select the function that is not an example of Poisson distribution:

a. The number of calls coming per minute into a call centre.
b. The number of soda cans sold per day in a convenience shop.
c. The number of cake slices eaten by a family in 3 hours.
d. The number of customers coming to a restaurant in 1 hour.

Q2. Which statement is correct?

a. When plotted in a log-log scale, the degree distribution of a social network should be in a horizontal line.
b. The power laws are scale free.
c. The degrees of nodes in a social network distribute evenly.
d. The degree distribution in a random graph follows power law.

Q3. A tree is a symmetric regular tree constructed starting from a central node of degree k as shown below. In this tree network, every node at distance d from the central node has degree k until we reach the nodes at distance P that have degree one and are called the leaves of the network. In this question, take $k = 3$. Does the network show small-world phenomena property? Justify your answer.

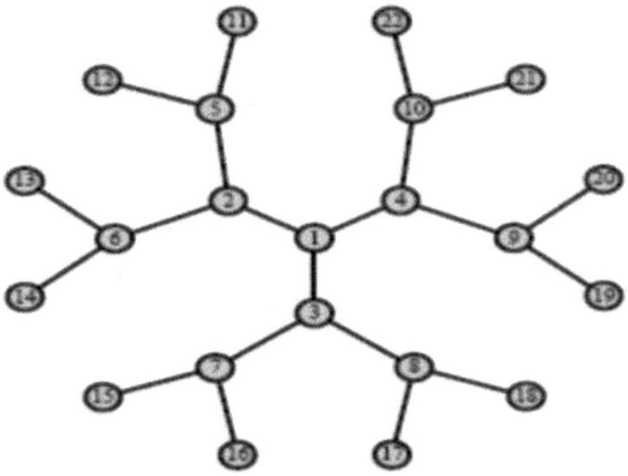

Q4. With proper examples, explain the robustness of random networks and scale-free networks with respect to epidemics.

Q5. A random network is constructed such that a pair of nodes are connected with a probability of 0.4. What is the expected number of possible edges if there are 100 nodes?

Q6. What is the difference between a scale-free network and a small-world network? Describe what is meant by 'six degrees of separation'. Explain the difference between close friendships and acquaintance in a social structure, and how these might impact six degrees of separation.

Q7. Given a network being generated by the 'rich get richer' phenomenon, the figure below shows the snapshot of the network at time t. A new node u enters the network at time $t + 1$ and makes an edge with one of the existing nodes. What is the probability that u will make an edge with C?

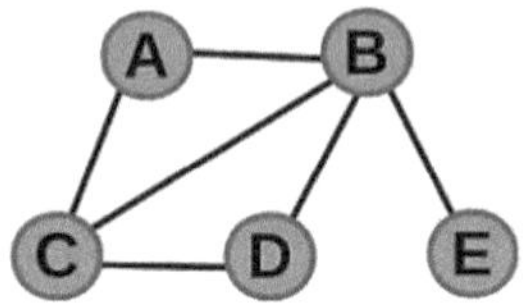

REFERENCES

1. Easley, D., & Kleinberg, J. (2010). *Networks, Crowds, and Markets: Reasoning About a Highly Connected World.* Cambridge University Press.
2. Barker, J. (1999). Power law. https://www.ncnr.nist.gov/resources/sansmodels/PowerLaw.html.
3. Newman, M. (2010). *Networks: An Introduction.* Oxford University Press.
4. Fabian, M. Erdös–Rényi model: Definition. https://maelfabien.github.io/machinelearning/graph_2/#definition
5. Social networks: Theories and applications. http://leonidzhukov.net/hse/2014/socialnetworks/
6. Zhukov, L. E. (2015). *Mathematical models of networks.* https://pdfs.semanticscholar.org/9b96/2fb751f46b23a06b9e3114c974e88f0726ce.pdf
7. Hart, M. G., Ypma, R. J. F., Romero-Garcia, R., Price, S. J., & Suckling, J. (2016). Graph theory analysis of complex brain networks: New concepts in brain mapping applied to neurosurgery. *Journal of Neurosurgery* 124(6), 1665–1678.
8. Wasserman, S., & Faust, K. (1994). *Social Network Analysis. Methods and Applications.* Cambridge University Press.
9. Six degrees of separation. *WhatIs.com.* https://whatis.techtarget.com/definition/six-degrees-of-separation
10. Adomaitis, M. B. Theory of six degrees of separation. *Love to Know.* https://socialnetworking.lovetoknow.com/Theory_of_Six_Degrees_of_Separation

4 Network Centrality

Why do we need measures? Who are the central figures (influential individuals) in the network? What interaction patterns are common in friends? Who are the like-minded users and how can we find these similar individuals? To answer these and similar questions, one first needs to define measures for quantifying centrality, level of interactions and similarity, among others.

In this chapter, we will introduce the concept of centrality and the various measures that have been associated with this concept. We will show all the measurements with the help of some applications.

4.1 CENTRALITY MEASURES OVERVIEW

Centrality measures are a vital tool for understanding social networks or social graphs. There are various measures that are used to evaluate the centrality of the complex networks. Centrality matrices use graph theory to calculate the importance of any given node in a social network. They cut through noisy data, revealing parts of the network that need attention – but they all work differently. Each measure has its own definition of 'importance', so we need to understand how they work to find the best one for our graph visualization applications [1].

Thus, we can say, centrality tells us which nodes are important in a network based on the topological structure of the network (instead of just looking at the popularity of nodes), such as:

- How influential a person is within a social network?
- Which genes play a crucial role in regulating systems and processes?
- Infrastructure networks: if the node is removed, it would critically impede the functioning of the network.

Importance of the Centrality Measure: Knowing the structure of a network, we can calculate various useful quantities or measures that capture particular features of the network topology.

Finding out which is the most central node is important, it could help with:

- disseminating information in the network faster;
- stopping epidemics;
- protecting the network from breaking;
- evenly distributing centrality among nodes;
- determining which nodes are most 'central'?

Centrality Visualization: Centrality can have various meanings.

Different Centrality Measures: In this chapter, we will look at the following centrality measures, how they work and when to use them.

DOI: 10.1201/9781003088066-4

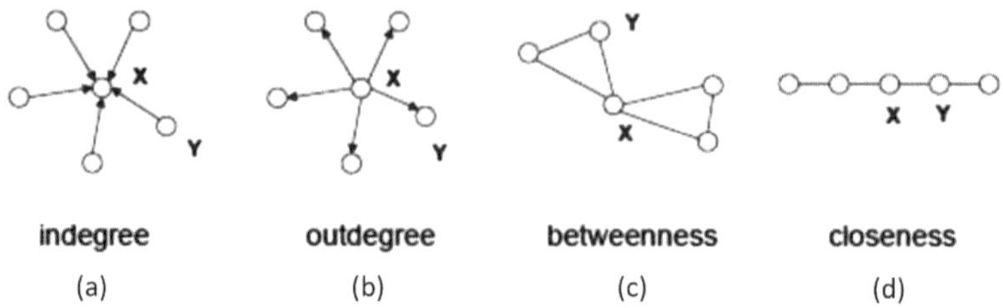

FIGURE 4.1 Centrality visualization

Centrality in terms of those who you are connected to:

1. Degree centrality
2. Eigenvector centrality
3. Katz centrality

Centrality in terms of how you connect others (information broker):

4. Betweenness centrality

Centrality in terms of how fast you can reach others:

5. Closeness centrality

4.2 DEGREE CENTRALITY

This measures the number of vertices adjacent to a vertex (degree). Degree centrality assigns an importance score based simply on the number of links held by each node. It tells us how many direct, 'one hop' connections each node has to other nodes in the network. It can be used for finding very connected individuals, popular individuals, individuals who are likely to hold most information or individuals who can quickly connect with the wider network [2].

Degree centrality is the simplest measure of node connectivity. Sometimes, it's useful to look at indegree (number of inbound links) and outdegree (number of outbound links) as distinct measures, for example when looking at transactional data or account activity. Degree centrality can be measured when the number of connections is the best centrality measure. For example:

- People who will do favours for you
- People you can talk to/have a beer with
- People you can talk to (influence set, information access, ...)
- Influence of an article in terms of citations (using indegree)

Normalized Degree centrality: This is a calculated degree of a node divided by the maximum possible degree of the network, i.e., $(N - 1)$.

$$\text{Normalized degree centrality} = \frac{\text{Degree of node } i}{N - 1} \tag{4.1}$$

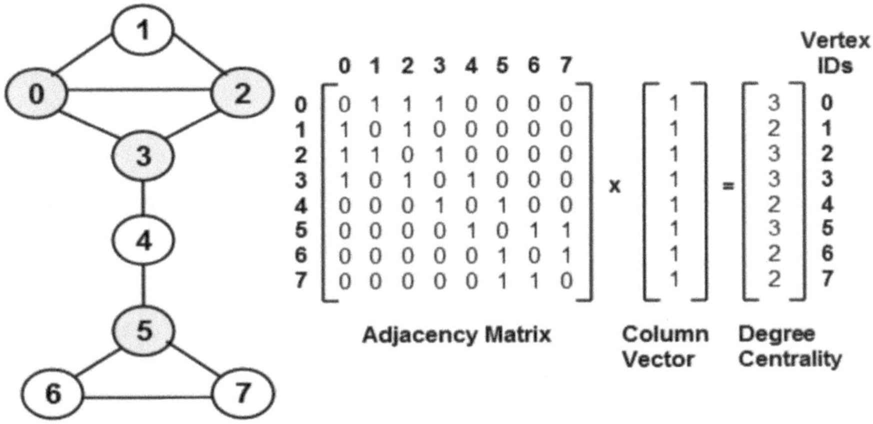

FIGURE 4.2 Degree centrality measure [1]

Where, N is total number of nodes in the network.

For example:

Consider Figure 4.3(a):

$$\frac{2}{7-1} = \frac{1}{3} = 0.33, \frac{3}{7-1} = \frac{1}{2} = 0.50$$

Consider Figure 4.3(b):

$$\frac{1}{6-1} = \frac{1}{5} = 0.2, \frac{5}{6-1} = \frac{1}{1} = 1$$

Consider Figure 4.3(c):

$$\frac{1}{5-1} = \frac{1}{4} = 0.25, \frac{2}{5-1} = \frac{2}{4} = \frac{1}{2} = 0.50$$

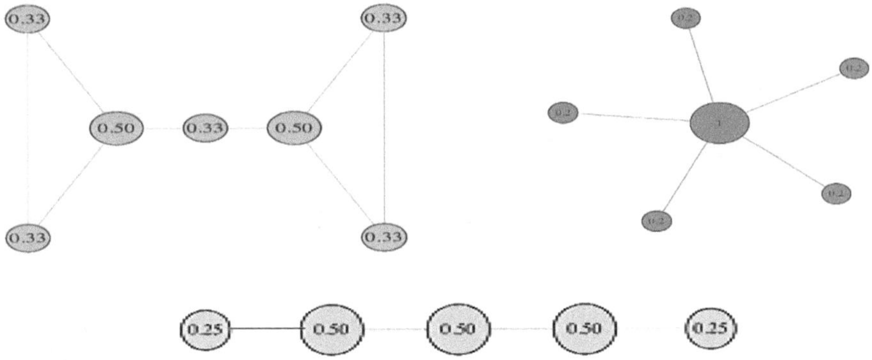

FIGURE 4.3 Normalized degree centrality

Weakness: It is very likely that more than one vertex has the same degree and it is not possible to uniquely rank the vertices.

Freeman's general formula for centralization:

$$C_D = \frac{\sum_{i=1}^{g}\left[C_D\left(n^*\right) - C_D(i)\right]}{\left[(N-1)(N-2)\right]} \tag{4.2}$$

Where, $C_D\left(n^*\right)$ = Max value of degree centrality in the network.

Example: consider Figure 4.3(a):

$$C_D = \frac{(3-2)+(3-2)+(3-3)+(3-2)+(3-3)+(3-2)+(3-2)}{(7-1)(7-2)} = \frac{5}{30} = \frac{1}{6} = 0.167$$

Consider Figure 4.3(b):

$$C_D = \frac{(5-1)+(5-1)+(5-1)+(5-1)+(5-1)+(5-5)}{(6-1)(6-2)} = \frac{20}{20} = 1$$

Consider Figure 4.3(c):

$$C_D = \frac{(2-1)+(2-2)+(2-2)+(2-2)+(2-1)}{(5-1)(5-2)} = \frac{2}{12} = \frac{1}{6} = 0.167$$

Note that you can use other metrics, for example, the Gini coefficient or standard deviation for centralization.

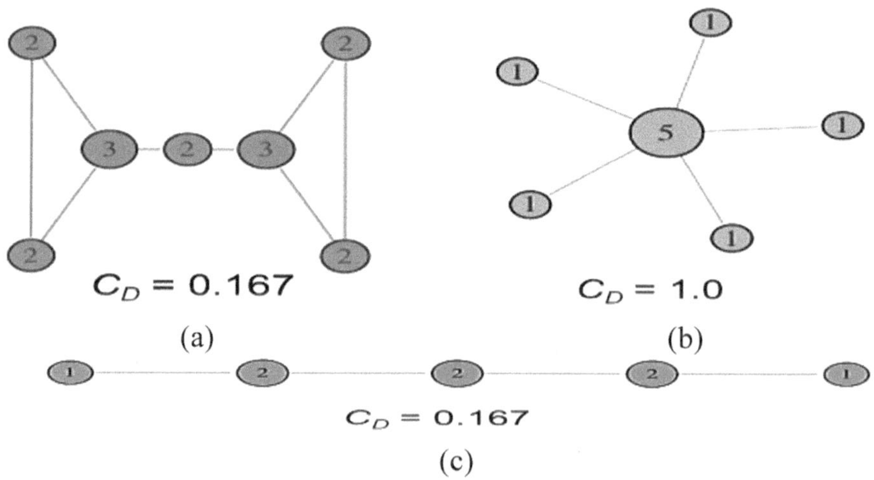

FIGURE 4.4 Freeman's degree centralization

4.3 EIGENVECTOR CENTRALITY

Having more friends does not by itself guarantee that someone is more important. Having more important friends provides a stronger signal. Eigenvector centrality incorporates the importance of the neighbours (undirected). For directed graphs, we can use incoming or outgoing edges.

Like degree centrality, eigenvector centrality [3] measures a node's influence based on the number of links it has to other nodes in the network. Eigenvector centrality then goes a step further by also taking into account how well connected a node is, and how many links their connections have, and so on through the network. By calculating the extended connections of a node, eigenvector centrality can identify nodes with influence over the whole network, not just those directly connected to it. Eigenvector centrality is a good 'all-round' SNA score, handy for understanding human social networks, but also for understanding networks like malware propagation. Tools can calculate each node's eigenvector centrality by converging on an eigenvector using the power iteration method.

Let us assume the eigenvector centrality of a node is $C_e(v_i)$. We would like $C_e(v_i)$ to be higher when important neighbours (node v_j with higher $C_e(v_i)$ point to us. For incoming neighbours $A_{j,i} = 1$

We can assume that v_i' scentrality is the summation of its neighbour's centralities

$$C_e(v_i) = \sum_{j=1}^{n} A_{j,i} C_e(v_j) \tag{4.3}$$

Normalized eigenvector centrality:

$$C_e(v_i) = \frac{1}{\lambda} \sum_{j=1}^{n} A_{j,i} C_e(v_j) \tag{4.4}$$

where λ is some fixed constant (eigenvalue).

Let $C_e = \left(C_e(v_1), C_e(v_2), \ldots, C_e(v_n)\right)^T \rightarrow \lambda C_e = A^T C_e$

This means that C_e is an eigenvector of adjacency matrix A^T (or A when undirected) and λ is the corresponding eigenvalue. Which eigenvalue–eigenvector pair should we choose to be based on Perron–Frobenius theorem?

Let $A = (a_{ij})$ be an $n \ x \ n$ symmetric matrix with elements $a_{ij} > 0$ and let λ be the largest eigenvalue. Then (i) $\lambda > 0$, (ii) there exists a corresponding eigenvector (x_j) with every entry $x_j > 0$, (iii) λ is non-degenerate and (iv) if μ is any other eigenvalue, $\lambda > |\mu|$

To compute eigenvector centrality of A,

- We compute the eigenvalues of A
- Select the largest eigenvalue λ.
- The corresponding eigenvector of λ is C_e.
- Based on the Perron–Frobenius theorem, all the components of C_e will be positive.
- The components of C_e are the eigenvector centralities for the graph.

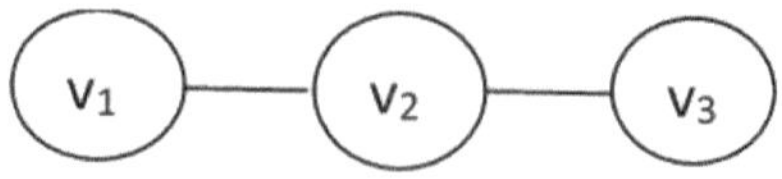

FIGURE 4.5 Example of eigenvector centrality

Eigenvector Centrality Example:

$$\lambda C_e = AC_e$$

$$A = \begin{bmatrix} 0 & 1 & 0 \\ 1 & 0 & 1 \\ 0 & 1 & 0 \end{bmatrix}$$

$$(A - \lambda I)C_e = 0 \qquad C_e = \begin{bmatrix} u_1 & u_2 & u_3 \end{bmatrix}^T$$

$$\left(\begin{bmatrix} 0 & 1 & 0 \\ 1 & 0 & 1 \\ 0 & 1 & 0 \end{bmatrix} - \lambda I \right) \begin{bmatrix} u_1 \\ u_2 \\ u_3 \end{bmatrix} = \begin{bmatrix} 0 \\ 0 \\ 0 \end{bmatrix}$$

$$\left(\begin{bmatrix} 0-\lambda & 1 & 0 \\ 1 & 0-\lambda & 1 \\ 0 & 1 & 0-\lambda \end{bmatrix} \right) \begin{bmatrix} u_1 \\ u_2 \\ u_3 \end{bmatrix} = \begin{bmatrix} 0 \\ 0 \\ 0 \end{bmatrix}$$

$$\det(A - \lambda I) = \begin{vmatrix} 0-\lambda & 1 & 0 \\ 1 & 0-\lambda & 1 \\ 0 & 1 & 0-\lambda \end{vmatrix} = 0$$

$$(-\lambda)(\lambda^2 - 1) - 1(-\lambda) = 2\lambda - \lambda^3 = \lambda(2 - \lambda^2) = 0$$

Thus, eigenvalues are $\left(-\sqrt{2}, 0, +\sqrt{2}\right)$, and largest eigenvalue λ is $+\sqrt{2}$. Corresponding eigenvector (assuming C_e has norm 1)

$$\left(\begin{bmatrix} 0-\sqrt{2} & 1 & 0 \\ 1 & 0-\sqrt{2} & 1 \\ 0 & 1 & 0-\sqrt{2} \end{bmatrix} \right) \begin{bmatrix} u_1 \\ u_2 \\ u_3 \end{bmatrix} = \begin{bmatrix} 0 \\ 0 \\ 0 \end{bmatrix}$$

$$C_e = \begin{bmatrix} u_1 \\ u_2 \\ u_3 \end{bmatrix} = \begin{bmatrix} 1/2 \\ \sqrt{2}/2 \\ 1/2 \end{bmatrix}$$

4.4 KATZ CENTRALITY

A major problem with eigenvector centrality arises when it deals with directed graphs. Centrality only passes over outgoing edges and in special cases such as when a node is in a weakly connected component centrality becomes zero even though the node can have many edges connected to it. To resolve this problem, we add bias term β to the centrality values for all nodes.

$$C_{\text{Katz}}(v_i) = \alpha \sum_{j=1}^{n} A_{j,i} C_{\text{Katz}}(v_j) + \beta \tag{4.5}$$

$$C_e(v_i) = \frac{1}{\lambda} \sum_{j=1}^{n} A_{j,i} C_e(v_j) \text{ (eigenvector centrality)}$$

$$C_{\text{Katz}}(v_i) = \alpha \sum_{j=1}^{n} A_{j,i} C_{\text{Katz}}(v_j) + \beta \tag{4.6}$$

Where α is a controlling term and β is a bias term.

When $\alpha = 0$, the eigenvector centrality is removed and all nodes get the same centrality value β. As α gets larger, the effect of β is reduced.

Rewriting equation in a vector form:

$$\boldsymbol{C}_{\text{Katz}} = \alpha A^T \boldsymbol{C}_{\text{Katz}} + \beta \mathbf{1} \tag{4.7}$$

where $\mathbf{1}$ is vector of all 1's.

Katz centrality:

$$\boldsymbol{C}_{\text{Katz}} = \beta\left(\boldsymbol{I} - \alpha A^T\right).\mathbf{1} \tag{4.8}$$

For the matrix $\left(\boldsymbol{I} - \alpha A^T\right)$ to be invertible, we must have:

- $det\left(\boldsymbol{I} - \alpha A^T\right) \neq 0$
- By rearranging we get $det\left(A^T - \alpha^{-1} I\right) = 0$
- This is basically the characteristic equation, which first becomes zero when the largest eigenvalue equals α^{-1}

In practice, we select $\alpha < 1/\lambda$, where λ is the largest eigenvalue of A^T. The largest eigenvalue is easier to compute (power method).

Katz Centrality Example:

$$A = \begin{bmatrix} 0 & 1 & 1 & 1 & 0 \\ 1 & 0 & 1 & 1 & 1 \\ 1 & 1 & 0 & 1 & 1 \\ 1 & 1 & 1 & 0 & 0 \\ 0 & 1 & 1 & 0 & 0 \end{bmatrix} = A^T$$

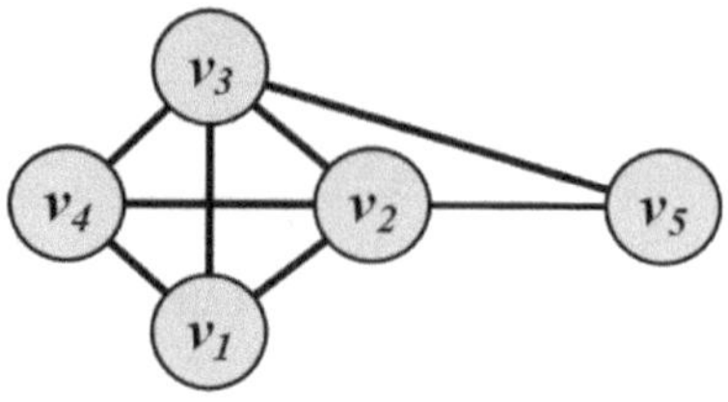

FIGURE 4.6 Example of Katz centrality

The eigenvalues are −1.68, −1.0, 0.35, 3.32. We assume $\alpha = 0.25 < \frac{1}{3.32}$ and $\beta = 0.2$

$$C_{\text{Katz}} = \beta\left(\boldsymbol{I} - \alpha A^T\right).\mathbf{1} = \begin{bmatrix} 1.14 \\ 1.31 \\ 1.31 \\ 1.14 \\ 0.85 \end{bmatrix}$$

Thus, v_2 *and* v_3 and the most important nodes.

The problem with Katz centrality: In directed graphs, once a node becomes an authority (high centrality), it passes all its centrality along all of its out-links. This is less desirable since not everyone known by a well-known person is well known.

We can divide the value of passed centrality by the number of outgoing links (i.e., outdegree of that node). Each connected neighbour gets a fraction of the source node's centrality. Now, consider when the degree isn't everything. In what ways does the degree fail to capture centrality [4] in the networks given in Figure 4.3. It fails in the ability to broker between groups and the likelihood that information originating anywhere in the network reaches you.

4.5 BETWEENNESS CENTRALITY

Another way of looking at centrality is by considering how important nodes are in connecting other nodes:

$$C_b\left(v_i\right) = \sum_{s \neq t \neq v} \frac{\sigma_{st}\left(v_i\right)}{\sigma_{st}} \tag{4.9}$$

σ_{st} the number of shortest paths from vertex s to t, a.k.a. information pathways.
$\sigma_{st}\left(v_i\right)$ the number of shortest paths from s to t that pass-through v_i

Betweenness centrality measures the number of times a node lies on the shortest path between other nodes. This measure shows which nodes are 'bridges' between nodes in a network. It does this by identifying all the shortest paths and then counting how many times each node falls on one. It is used for finding the individuals who influence the flow around a system. Betweenness is useful for analyzing communication dynamics, but should be used with care [5]. A high betweenness count could indicate that someone holds authority over disparate clusters in a network, or just that they are on the periphery of both clusters.

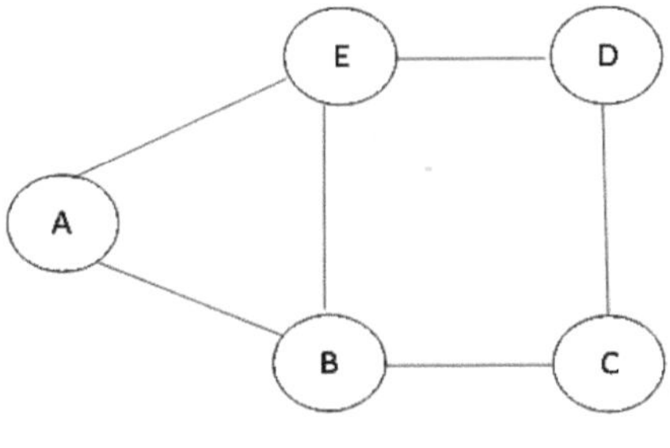

FIGURE 4.7 Example of betweenness centrality

Normalizing betweenness centrality: In the best case, node v_i is on all shortest paths from s *to* t, hence,

$$\frac{\sigma_{st}(v_i)}{\sigma_{st}} = 1$$

$$C_b(v_i) = \sum_{s \neq t \neq v} \frac{\sigma_{st}(v_i)}{\sigma_{st}} = \sum_{s \neq t \neq v} 1 = 2\binom{n-1}{2} = (n-1)(n-2).$$

Therefore, the maximum value is $2\binom{n-1}{2}$

$$C_b^{\text{norm}}(v_i) = \frac{C_b(v_i)}{[(n-1)(n-2)/2]} \tag{4.10}$$

Betweenness Centrality Example:

Now, find the shortest path from each node to every node and intermediate node in that path (see Table 4.1).

TABLE 4.1
Shortest Path from Source to Target Node

Source node (s)	Target node (t)	Intermediate nodes in the path (v)	Path
A	B		AB
A	C	B	ABC
A	D	E	AED
A	E		AE
B	C		BC
B	D	E or C	BCD/BED
B	E		BE
C	D		CD
C	E	B or D	CBE/CDE
D	E		DE

With this table, we can observe that node A is not an intermediate node in any path. Thus, the betweenness centrality of node A is 0: $C_b(A)=0$.

$$C_b(B)=\sum_{s\neq t\neq B}\frac{\sigma_{st}(B)}{\sigma_{st}}=\frac{1+0.5}{12}=\frac{1.5}{12}=0.125$$

As σ_{st} = total shortest path = 12

Normalized betweenness =

$$C_b^{\text{norm}}(B)=\frac{C_b(B)}{[(n-1)(n-2)/2]}=\frac{0.125}{\left[\frac{(5-1)(5-2)}{2}\right]}=\frac{0.125}{6}=0.0208$$

$$C_b^{\text{norm}}(C)=\frac{C_b(C)}{[(n-1)(n-2)/2]}=\frac{0.5/12}{\left[\frac{(5-1)(5-2)}{2}\right]}=\frac{0.042}{6}=0.0069$$

$$C_b^{\text{norm}}(D)=\frac{C_b(D)}{[(n-1)(n-2)/2]}=\frac{0.5/12}{\left[\frac{(5-1)(5-2)}{2}\right]}=\frac{0.042}{6}=0.0069$$

$$C_b^{\text{norm}}(E)=\frac{C_b(E)}{[(n-1)(n-2)/2]}=\frac{1.5/12}{\left[\frac{(5-1)(5-2)}{2}\right]}=\frac{0.125}{6}=0.0208$$

What if it's not so important to have many direct friends, or to be 'between' others but one still wants to be in the 'middle' of things, not too far from the centre?

4.6 CLOSENESS CENTRALITY

Closeness centrality scores each node based on their 'closeness' to all other nodes in the network. This measure calculates the shortest paths between all nodes, then assigns each node a score based on its sum of shortest paths. It is used for finding the individuals who are best placed to influence the entire network most quickly. Closeness centrality can help find good 'broadcasters', but in a highly-connected network, you will often find all nodes have a similar score. What may be more useful is using closeness to find influencers in a single cluster [6].

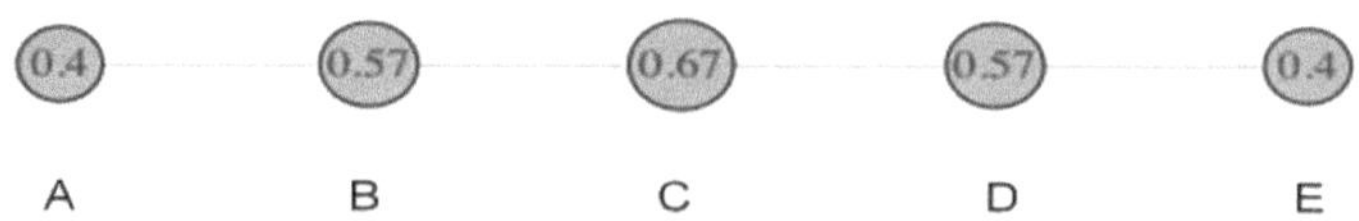

FIGURE 4.8 Example of the closeness centrality measure

The intuition is that influential and central nodes can quickly reach other nodes. These nodes should have a smaller average shortest path length to other nodes. Closeness is based on the length of the average shortest path between a vertex and all vertices in the graph. It depends on inverse distance to other vertices.

$$\text{Closeness centrality } C_c(i) = \left[\sum_{j=1}^{N} d(i,j)\right]^{-1} \tag{4.11}$$

Normalized closeness centrality:

$$C_c'(i) = (C_c(i)) \cdot (N-1) \tag{4.12}$$

Closeness Centrality Measures Example:

$$C_c'(A) = \left[\frac{\sum_{j=1}^{N} d(A,j)}{N-1}\right]^{-1} = \left[\frac{1+2+3+4}{4}\right]^{-1} = \left[\frac{10}{4}\right]^{-1} = 0.4$$

$$C_c'(B) = \left[\frac{\sum_{j=1}^{N} d(B,j)}{N-1}\right]^{-1} = \left[\frac{1+1+2+3}{4}\right]^{-1} = \left[\frac{7}{4}\right]^{-1} = [1.75]^{-1} = 0.57$$

$$C_c'(C) = \left[\frac{\sum_{j=1}^{N} d(C,j)}{N-1}\right]^{-1} = \left[\frac{2+1+1+2}{4}\right]^{-1} = \left[\frac{6}{4}\right]^{-1} = [1.5]^{-1} = 0.67$$

$$C_c'(D) = \left[\frac{\sum_{j=1}^{N} d(D,j)}{N-1}\right]^{-1} = \left[\frac{3+2+1+1}{4}\right]^{-1} = \left[\frac{7}{4}\right]^{-1} = [1.75]^{-1} = 0.57$$

$$C_c'(F) = \left[\frac{\sum_{j=1}^{N} d(F,j)}{N-1}\right]^{-1} = \left[\frac{4+3+2+1}{4}\right]^{-1} = \left[\frac{10}{4}\right]^{-1} = 0.4$$

In this chapter, we discussed measures for a social media network.

4.7 PROBLEMS FOR SELF-ASSESSMENT

Q1. For the undirected network shown below, compute the betweenness and closeness centrality for node A, B and D.

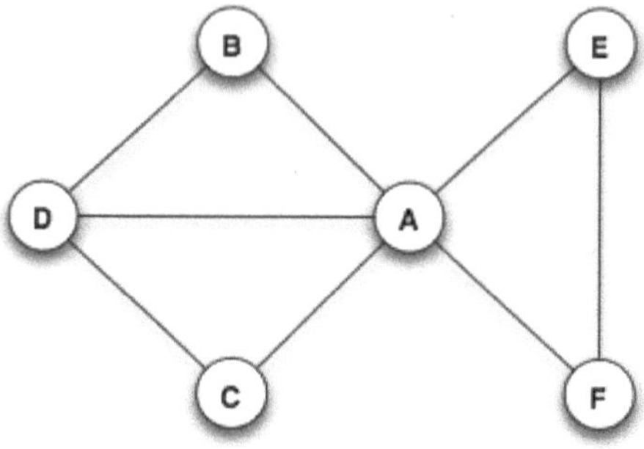

Q2. Consider the two networks below. For each graph, calculate (show all work):

a. Normalized degree centrality for each node
b. Normalized closeness centrality of each node
c. Normalized betweenness centrality of each node

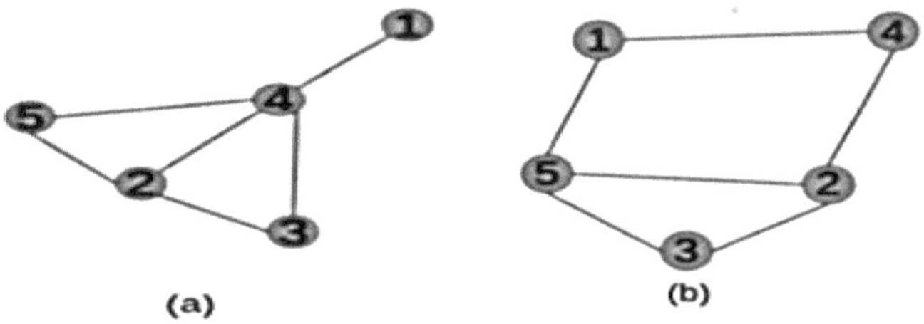

Q3. Consider the network below. What is the normalized degree centrality and normalized betweenness centrality of vertices C and F in the network?

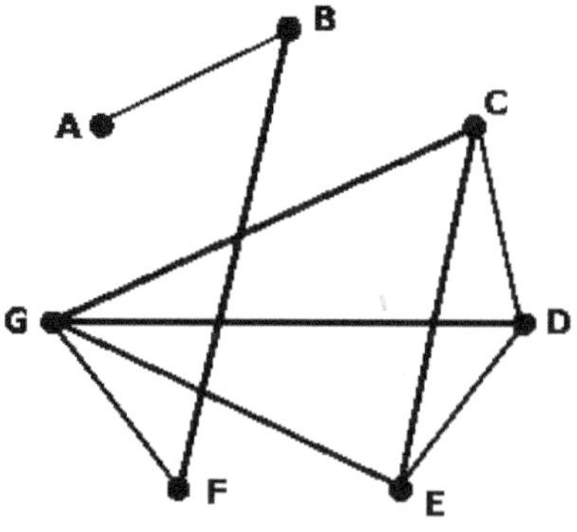

Q4. Calculate the betweenness centrality, closeness centrality, eigenvector centrality and Katz centrality of all the nodes in the below network.

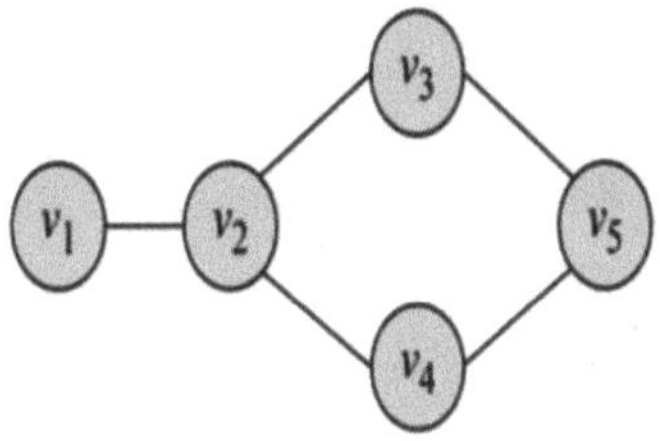

REFERENCES

1. Meghanathan, M. (2019). *Centrality metrics.* https://www.jsums.edu/nmeghanathan/files/2019/08/CSC641-Fall2019-Module-3-Centrality.pdf
2. Social media mining. *Network measures.* https://docplayer.net/9479548-Social-media-mining-network-measures.html
3. Disney, A. (2020, Jan 2). *Social network analysis 101: Centrality measures explained. Cambridge Intelligence.* https://cambridge-intelligence.com/keylines-faqs-social-network-analysis/
4. Hanneman, R. A., & Riddle, M. *Introduction to social networks: Centrality and power.* https://faculty.ucr.edu/~hanneman/nettext/C10_Centrality.html
5. Mascolo, C. *Centrality measures.* https://www.cl.cam.ac.uk/teaching/1314/L109/stna-lecture3.pdf
6. Zafarani, R., Abbasi, M. A., & Liu, H. (2014). Network measures, *Social Media Mining: An Introduction.* Cambridge University Press. https://cse.iitkgp.ac.in/~bivasm/cnt_notes/centrality.pdf

5 Link Analysis

Link analysis is a data analysis technique used in network theory that is used to evaluate the relationships or connections between network nodes. These relationships can be between various types of objects (nodes), including people, organizations and even transactions.

Link analysis is essentially a kind of knowledge discovery that can be used to visualize data to allow for better analysis, especially in the context of links, whether web links or relationship links between people or between different entities. Link analysis is often used in search engine optimization as well as in intelligence, in security analysis and in market and medical research [1].

Link analysis is literally about analyzing the links between objects, whether they are physical, digital or relational. This requires diligent data gathering. For example, in the case of a website where all of the links and backlinks that are present must be analyzed, a tool has to sift through all of the HTML codes and various scripts in the page and then follow all the links it finds in order to determine what sort of links are present and whether they are active or dead. This information can be very important for search engine optimization because it allows the analyst to determine whether the search engine is actually able to find and index the website [1].

Link analysis has three primary purposes:

- Find matches for known patterns of interests between linked objects.
- Find anomalies by detecting violated known patterns.
- Find new patterns of interest (for example, in social networking and marketing and business intelligence).

5.1 LINK ANALYSIS IN WEB MINING

Hyperlink analysis by itself is a part of a bigger research domain – web mining – which can be described as the process of applying data mining techniques to extract useful information from web data. The field of web mining can be broadly divided into three distinct categories, according to the kinds of data to be mined [2]:

1. Web Content Mining: Web content mining is the process of extracting useful information from the contents of web documents. Content data is the collection of facts a web page is designed to contain. It may consist of text, images, audio, video or structured records such as lists and tables.
2. Web Structure Mining (WSM): The structure of a typical web graph consists of web pages as nodes, and hyperlinks as edges connecting two related pages. WSM is used to find out the relation between different web pages by processing the structure of web. WSM can be performed at two levels:
   a. Document structure analysis: deals with the structure of a document such as the Document Object Model.
   b. Link type analysis: deals with links that may be inter-document or intra-document.

DOI: 10.1201/9781003088066-5

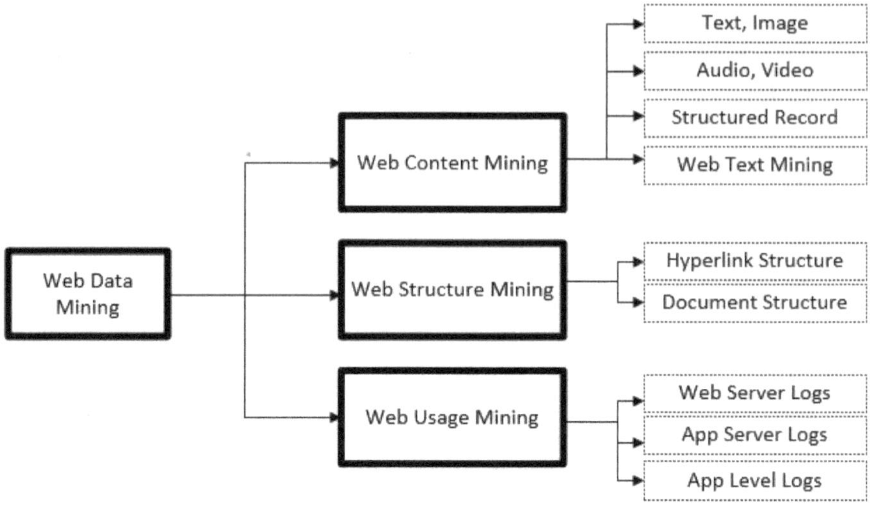

FIGURE 5.1 Taxonomy of web mining

3. Web Usage Mining: Web usage mining is the application of data mining techniques to discover interesting usage patterns from web usage data, in order to understand and better serve the needs of web-based applications. Usage data captures the identity or origin of web users along with their browsing behaviour at a website. Web usage mining itself can be classified further depending on the kind of usage data considered: web server data, application server data and application-level data.

5.2 RANKING ALGORITHMS

The web page ranking algorithms rank the search results depending upon their relevance to the search query. For these, algorithms rank the search results in descending order of relevance to the query string being searched. A web page's ranking for a specific query depends on factors such as its relevance to the words and concepts in the query, its overall link popularity and so on. There are two categories of these algorithms: text-based ranking and link-based ranking.

1. **Text Based Ranking:** The ranking scheme used in the conventional search engines is purely text-based (i.e., the pages are ranked based on their textual content, which seems to be logical). In such schemes, the factors that influence the rank of a page are:
    - Number of matched terms.
    - Location of terms in document.
    - Frequency of terms in document.
2. **Link-Based Ranking Algorithm:** This algorithm views the web as a directed graph where the web pages form the nodes and the hyperlinks

between the web pages form the directed edges between these nodes. Link-based ranking algorithms propagate page importance through links. Two most influential hyperlink-based search algorithms are:

- HITS (Hyperlink-Induced Topic Search)
- Page-ranking algorithm

5.3 HYPERLINK-INDUCED TOPIC SEARCH (HITS)

HITS was developed by Jon Kleinberg in 1998 to rate web pages. Ranking of the web page is decided by analyzing its textual contents against a given query. When the user issues a search query, HITS first expands the list of relevant pages returned by a search engine and then produces two rankings of the expanded set of pages: **authority ranking** and **hub ranking**. In this algorithm, a web page is named as authority if the web page is pointed by many hyperlinks and a web page is named as a hub if the page points to various hyperlinks.

Hubs and Authorities: Given a query to a search engine, the set of highly relevant web pages are called roots. They are potential **authorities**. Pages that are not very relevant but point to pages in the root are called **hubs** [3].

Thus, an authority is a page that many hubs link to, whereas a hub is a page that links to many authorities.

HITS Algorithm: In the HITS algorithm, the first step is to retrieve the most relevant pages to the search query. This set is called the root set and can be obtained by taking the top pages returned by a text-based search algorithm. Then a base set is generated by augmenting the root set with all the web pages that are linked from it and some of the pages that link to it. The web pages in the base set and all hyperlinks among those pages form a focused subgraph. The HITS computation is performed only on this focused subgraph. The algorithm performs a series of iterations, each consisting of two basic steps: **authority update and hub update.** The hub score and authority score for a node is calculated with the following algorithm [4]:

1. Start with each node having a hub score and authority score of 1.
2. Run the authority update rule.
3. Run the hub update rule.
4. Normalize the values by dividing each hub score by square root of the sum of the squares of all hub scores, and dividing each authority score by square root of the sum of the squares of all authority scores.
5. Repeat from the second step as necessary.

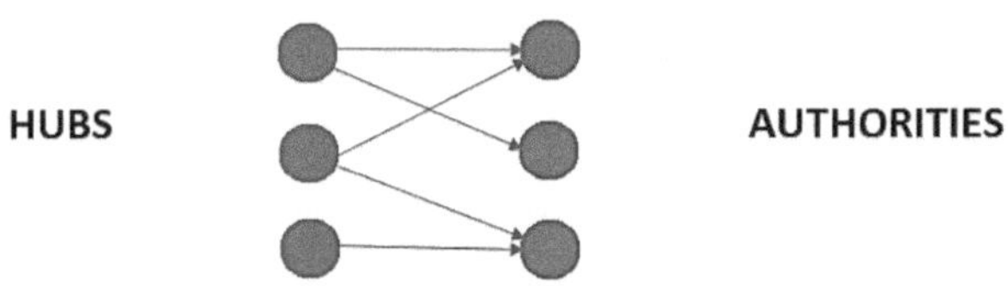

FIGURE 5.2 Example of hubs and authorities

Update Rules: To begin with for all p $\text{auth}(p)=1$ and $\text{hub}(p)=1$ where p is the page to be rank. A $k-$step application of the hub–authority algorithm entails applying for k times first the Authority Update Rule and then the Hub Update Rule.

- **Authority Update Rule:**

$$\forall p, \text{ we update auth}(p) \text{ to be the summation}:$$

$$\text{auth}(p)=\sum_{i=1}^{n}\text{hub}(i) \tag{5.1}$$

where n is the total number of pages connected to p and i is a page connected to p. That is, the authority score of a page is the sum of all the hub scores of pages that point to it.

- **Hub Update Rule:**

$$\forall p, \text{ we update hub}(p) \text{ to be the summation}:$$

$$\text{hub}(p)=\sum_{i=1}^{n}\text{auth}(i) \tag{5.2}$$

where n is the total number of pages p connects to and i is a page which p connects to. Thus, a page's hub score is the sum of the authority scores of all its linking pages

Normalization: The final hub–authority scores of nodes are determined after infinite repetitions of the algorithm. As directly and iteratively applying the Hub Update Rule and Authority Update Rule leads to diverging values, it is necessary to normalize the matrix after every iteration. By dividing each authority value by the square root of the sum of the squares of all authority values, and dividing each hub value by the square root of the sum of the squares of all hub values, the values obtained from this process will eventually converge.

Advantages of HITS:

- HITS scores due to its ability to rank pages according to the query string, resulting in relevant authority and hub pages.
- The ranking may also be combined with other information retrieval-based rankings.
- HITS is sensitive to user query (as compared to PageRank).
- Important pages are obtained on basis of calculated authority and hubs value.
- HITS is a general algorithm for calculating authority and hubs in order to rank the retrieved data.
- HITS induces web graph by finding sets of pages with a search on a given query string.
- Results demonstrate that HITS calculates authority nodes and hub-ness correctly

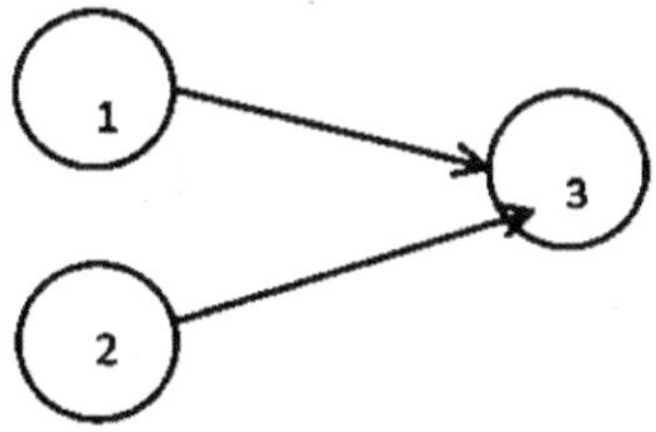

FIGURE 5.3 Example 1: Graph of the HITS algorithm

Disadvantages of HITS

- Query time cost
- Irrelevant authorities
- Irrelevant hubs
- Mutually reinforcing relationships between hosts
- Topic drift of less feasibility
- Less feasibility

HITS example:

$$A = \begin{pmatrix} 0 & 0 & 1 \\ 0 & 0 & 1 \\ 0 & 0 & 0 \end{pmatrix}, A^T = \begin{pmatrix} 0 & 0 & 0 \\ 0 & 0 & 0 \\ 1 & 1 & 0 \end{pmatrix}$$

Assume the initial hub weight vector is $u = \begin{pmatrix} 1 \\ 1 \\ 1 \end{pmatrix}$

Compute the authority weight vector $v = A^T \times u$

$$\begin{pmatrix} 0 & 0 & 0 \\ 0 & 0 & 0 \\ 1 & 1 & 0 \end{pmatrix} \times \begin{pmatrix} 1 \\ 1 \\ 1 \end{pmatrix} = \begin{pmatrix} 0 \\ 0 \\ 2 \end{pmatrix}$$

$$V = 0^2 + 0^2 + 2^2 = 4$$

Then update the hub weight:

$$u = A \times v$$

$$\begin{pmatrix} 0 & 0 & 1 \\ 0 & 0 & 1 \\ 0 & 0 & 0 \end{pmatrix} \times \begin{pmatrix} 0 \\ 0 \\ 2 \end{pmatrix} = \begin{pmatrix} 2 \\ 2 \\ 0 \end{pmatrix}$$

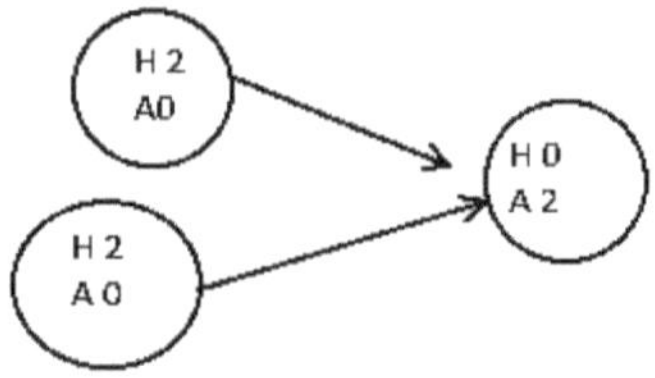

FIGURE 5.4 Graph with hub and authority scores

Compare hub score of each node with authority score of each node:

$$\text{Node 1 is hub as } 2 > 0$$

$$\text{Node 2 is hub as } 2 > 0$$

$$\text{Node 3 is authority as } 0 < 2$$

Example 2: Calculate the hubs and authority score using hits algorithm for $k = 3$

The adjacency matrix is:

	$N1$	$N2$	$N3$	$N4$
$N1$	0	1	1	1
$N2$	0	0	1	1
$N3$	1	0	0	1
$N4$	0	0	0	1

Rank using outdegree and indegree:

	Outdegree(Hub)	Indegree(Authority)
$N1$	3	1
$N2$	2	1
$N3$	2	2
$N4$	1	4

$$\text{Hubs} — N1,\ N2,\ N3\ (\text{Tie}),\ N4$$

$$\text{Authority} — N4,\ N3,\ N2,\ N1\ (\text{Tie})$$

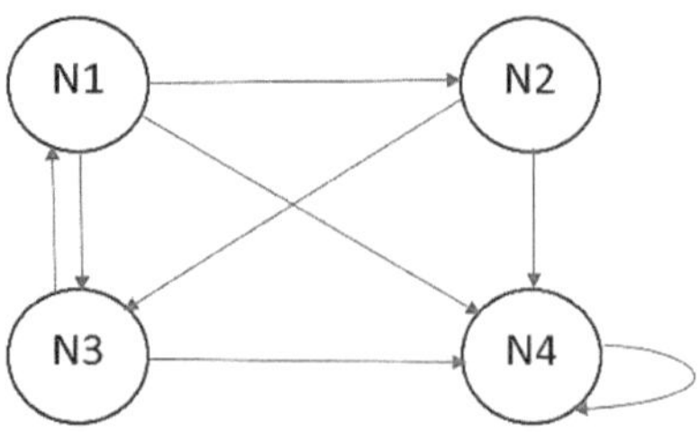

FIGURE 5.5 Graph of the HITS algorithm

Rank using HITS:

$$A = \begin{array}{c|cccc} & N1 & N2 & N3 & N4 \\ N1 & 0 & 1 & 1 & 1 \\ N2 & 0 & 0 & 1 & 1 \\ N3 & 1 & 0 & 0 & 1 \\ N4 & 0 & 0 & 0 & 1 \end{array}$$

$$\text{Transpose of } A^T = \begin{array}{c|cccc} & N1 & N2 & N3 & N4 \\ N1 & 0 & 0 & 1 & 0 \\ N2 & 1 & 0 & 0 & 0 \\ N3 & 1 & 1 & 0 & 0 \\ N4 & 1 & 1 & 1 & 1 \end{array}$$

Assuming, initial hub weight vector u as 1
Authority weight vector, $v = A^T \times u$

$$v = \begin{pmatrix} 0 & 0 & 1 & 0 \\ 1 & 0 & 0 & 0 \\ 1 & 1 & 0 & 0 \\ 1 & 1 & 1 & 1 \end{pmatrix} \times \begin{pmatrix} 1 \\ 1 \\ 1 \\ 1 \end{pmatrix} = \begin{pmatrix} 1 \\ 1 \\ 2 \\ 4 \end{pmatrix}$$

Updated hub weight vector $u = A \times v$

$$u = \begin{pmatrix} 0 & 1 & 1 & 1 \\ 0 & 0 & 1 & 1 \\ 1 & 0 & 0 & 1 \\ 0 & 0 & 0 & 1 \end{pmatrix} \times \begin{pmatrix} 1 \\ 1 \\ 2 \\ 4 \end{pmatrix} = \begin{pmatrix} 7 \\ 6 \\ 5 \\ 4 \end{pmatrix}$$

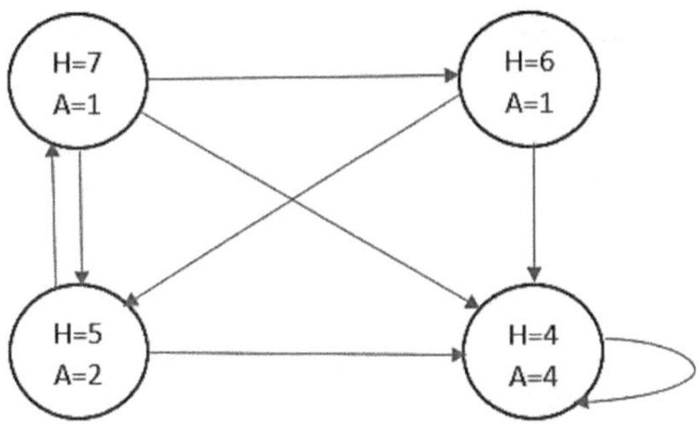

FIGURE 5.6 Graph with hub and authority scores after iteration 1

Now we create a graph again for $k = 1$

Node	Hub	Authority
$N1$	7	1
$N2$	6	1
$N3$	5	2
$N4$	4	4

$$\text{HUB}: N1,\ N2,\ N3,\ N4$$

$$\text{AUTHORITY}: N4,\ N3,\ N1,\ N2\ \{Tie\}$$

Then, calculate a new authority from $k = 1$:

$$v_1 = 1^2 + 1^2 + 2^2 + 4^2 = 1 + 1 + 4 + 16 = 22$$

$$v_1 = \frac{1}{\sqrt{22}}, \frac{1}{\sqrt{22}}, \frac{2}{\sqrt{22}}, \frac{4}{\sqrt{22}}$$

$$v_1 = 0.213\ (\text{N1}),\ 0.213(\text{N2}),\ 0.426(\text{N3}),\ 0.853(\text{N4})$$

Similarly, calculate a new hub from $k = 1$:

$$u_1 = 7^2 + 6^2 + 5^2 + 4^2 = 49 + 36 + 25 + 16 = 126$$

$$u_1 = \frac{7}{\sqrt{126}}, \frac{6}{\sqrt{126}}, \frac{5}{\sqrt{126}}, \frac{4}{\sqrt{126}}$$

$$u_1 = 0.623\ (\text{N1}),\ 0.535(\text{N2}),\ 0.445(\text{N3}),\ 0.356(\text{N4})$$

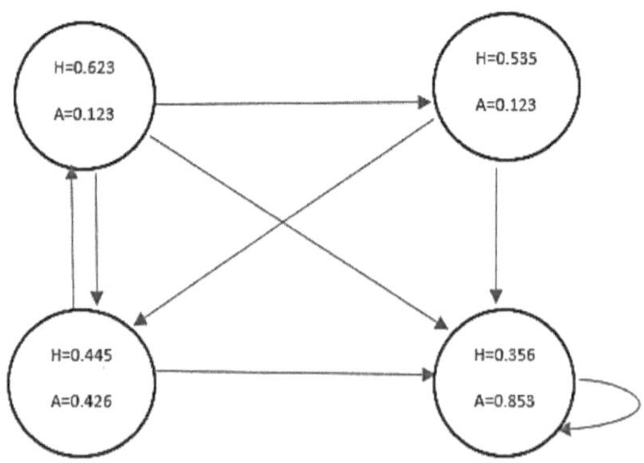

FIGURE 5.7 Graph with hub and authority scores after iteration 2

The diagram for $k = 2$, will be:

Nodes	Hubs (U1)	Authority (V1)
N1	0.623	0.213
N2	0.535	0.213
N3	0.445	0.426
N4	0.356	0.853

$$\text{HUBS}: N1, N2, N3, N4$$

$$\text{AUTHORITIES}: N4, N3, N2, N1\ (\text{TIE})$$

Then, calculate a new authority from $k = 2$:

$$v_2 = 0.2132 + 0.2132 + 0.4262 + 0.8532 = 0.999$$

$$v_2 = 0.213/\sqrt{0.999} + 0.213/\sqrt{0.999} + 0.426/\sqrt{0.999} + 0.853/\sqrt{0.999}$$

$$v_2 = 0.213 + 0.213 + 0.426 + 0.853$$

similarly, calculate a new hub from $k = 2$:

$$u_2 = 0.6232 + 0.5352 + 0.4452 + 0.3562 = 0.999$$

$$u_2 = 0.623/\sqrt{0.999} + 0.535/\sqrt{0.999} + 0.445/\sqrt{0.999} + 0.356/\sqrt{0.999}$$

$$u_2 = 0.623 + 0.535 + 0.445 + 0.356$$

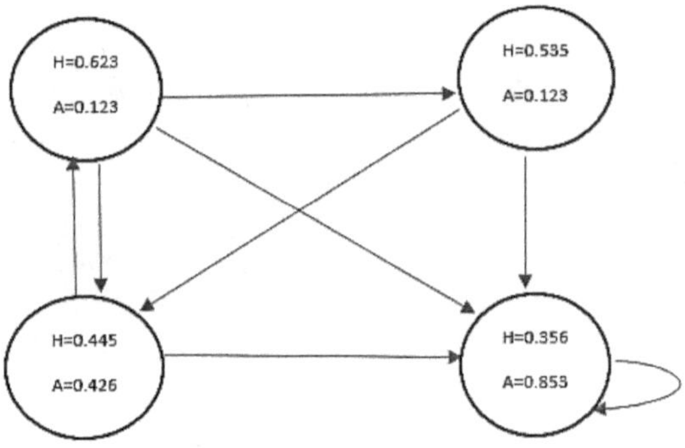

FIGURE 5.8 Graph with hub and authority scores after iteration 3

The diagram for $k = 3$, will be:

Nodes	Hubs (U1)	Authority (V1)
N1	0.623	0.213
N2	0.535	0.213
N3	0.445	0.426
N4	0.356	0.853

So, $k = 2 \approx k = 3$

Hub and authority scores have come to a consistent value. And when we reach this statement, we stop.

5.4 PAGERANK ALGORITHM

The PageRank algorithm was originally developed at Stanford University by Larry Page and Sergey Brin in 1996 as part of a research project about a new search engine. The algorithm assigns a numerical weight or rank to each page of a hyperlinked set of documents with the purpose of measuring its relative importance within the set. The PageRank algorithm utilizes link structure of the web pages. This algorithm is query independent and it operates on the whole web and assigns a PageRank to every web page. PageRank utilizes the back link in deciding the rank score.

$$PR(u) = \sum_{v \in Bu} \frac{PR(v)}{L(v)} \tag{5.3}$$

PageRank value for a web page u is dependent on the PageRank values for each web page v out of the set Bu (this set contains all pages linking to web page u), divided by the number $L(v)$ of links from page v. Ranking of web pages by the Google search engine was initially determined by three factors:

- Page specific factors
- Anchor text of inbound links
- PageRank

Algorithm: The PageRank algorithm does not rank the whole website, but it's determined for each page individually. Furthermore, the PageRank of page A is recursively defined by the PageRank of those pages which link to page A. PageRank is a probability distribution used to represent the likelihood that a person randomly clicking on links will arrive at any particular page. Thus, a PageRank of 0.5 means there is a 50% chance that a person clicking on a random link will be directed to the document with the 0.5 PageRank. Brin and Page described the PageRank formula as [5]

$$PR(p_i) = \frac{1-d}{N} + d \sum_{p_j \in M(p_i)} \frac{PR(p_j)}{L(p_j)} \tag{5.4}$$

where $p_1, p_2, \dots, p_N$ are the pages under consideration, $M(p_i)$ is the set of pages that link to p_i, $L(p_j)$ is the number of outbound links on page p_j, N is the total number of pages and d is the damping factor.

Damping Factor: The PageRank theory holds that any imaginary surfer who is randomly clicking on links will eventually stop clicking. The probability, at any step, that the person will continue is called a damping factor d. The damping factor can be set to any value such that $0 < d < 1$, but nominally it is set around 0.85. The damping factor is subtracted from 1 and this term is then added to the product of the damping factor and the sum of the incoming PageRank scores.

Advantages of the PageRank Algorithm:

- **Less Query Time Cost:** PageRank has a clear advantage over the HITS algorithm because the query time cost of incorporating the pre-computed PageRank importance score for a page is low.
- **Less Susceptibility to Localized Links:** Furthermore, as PageRank is generated using the entire web graph, rather than a small subset, it is less susceptible to localized link spam.
- **More Efficient:** In contrast, PageRank computes a single measure of quality for a page at crawl time. This measure is then combined with a traditional information retrieval score at query time. Compared with HITS, this has the advantage of much greater efficiency.
- **Feasibility:** As compared to the HITS algorithm, the PageRank algorithm is more feasible in today's scenario since it performs computations at crawl time rather than query time.

Disadvantages of the PageRank Algorithm:

- **Rank Sinks:** The rank sinks problem occurs when in a network pages get in infinite link cycles.
- **Spider Traps:** Another problem in PageRank is spider traps. A group of pages is a spider trap if there are no links from within the group to outside the group.
- **Dangling Links:** This occurs when a page contains a link such that the hypertext points to a page with no outgoing links.
- **Dead Ends:** Dead ends are simply pages with no outgoing links.
- **No Out Edges:** PageRank doesn't handle pages with no out edges very well, because they decrease the PageRank overall.
- **Circular References:** If you have circle references in your website, then it will reduce your front page's PageRank.
- **Effect of Additional Pages:** If you add a web page to your website, it will increase your page's rank by 0.428. The problem with this method is that if you increase your front page's PageRank by adding additional pages, then the rank of your other pages will go down.

The solution is to swap links with websites which have high PageRank value. The easiest way to do this is to make a page with high PageRank and link it to your front page. The PageRank score of a page ignores whether or not the page is relevant to the query at hand.

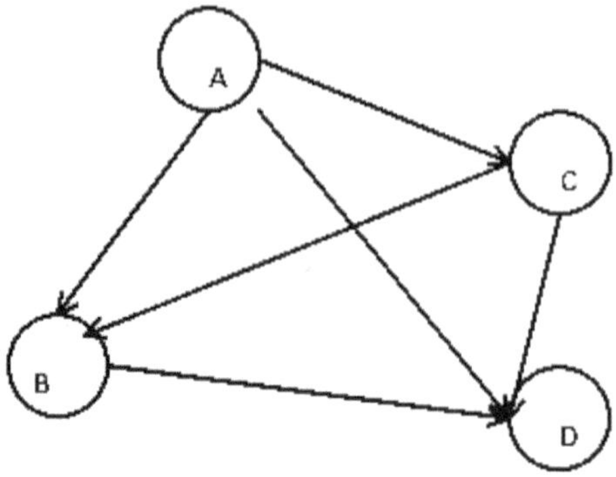

FIGURE 5.9 PageRank example: Graph 1

Page Rank Example:

$$PR(A)=(1-d)+d*\left(\frac{PR(T_1)}{C(T_1)}+\frac{PR(T_2)}{C(T_2)}+\ldots\right)$$

	Incoming	Outgoing
A	0	3
B	2	1
C	1	2
D	3	0

Initially, $PR(A)=PR(B)=PR(C)=PR(D)=1$

Iteration 1

$$PR\ (A)=(1-.8)+.8(0)=.2$$

$$PR\ (B)=(1-0.8)+0.8*\left(\frac{0.2}{3}+\frac{1}{2}\right)=0.2+0.453=0.653$$

$$PR(C)=0.2+0.8*\frac{0.2}{3}=0.253$$

$$PR\ (D)=0.2+0.8(0.2/3+0.653/1+0.253/2)$$

$$=0.2+0.8(.06666+0.653+0.1265)=0.8769$$

Iteration 2

$$PR\ (A)=0.2$$

$$PR\ (B)=0.2+0.8\left(\frac{0.2}{3}+\frac{0.253}{2}\right)=0.2+0.154=0.354$$

$$PR(C) = 0.2 + 0.8\left(\frac{0.2}{3}\right) = 0.253$$

$$PR\ (D) = 0.2 + 0.8\left(\frac{0.2}{3} + \frac{0.354}{1} + \frac{0.253}{2}\right) = 0.2 + 0.437 = 0.637$$

Iteration 3

$$PR\ (A) = 0.2$$

$$PR\ (B) = 0.2 + 0.8\left(\frac{0.2}{3} + \frac{0.253}{2}\right) = 0.2 + 0.154 = 0.354$$

$$PR(C) = 0.2 + 0.8\left(\frac{0.2}{3}\right) = 0.253$$

$$PR\ (D) = 0.2 + 0.8\left(\frac{0.2}{3} + \frac{0.354}{1} + \frac{0.253}{2}\right) = 0.2 + 0.437 = 0.637$$

Iteration 2 and iteration 3 scores have come to a consistent value. And when we reach this state, we stop.

Thus, $PR\ (A) = 0.2$; $PR\ (B) = 0.354$; $PR\ (C) = 0.253$; $PR\ (D) = 0.637$

Example 2:

The number of web pages $N = 3$

The damping parameter $d = 0.7$

$$PR\ (A) = (1 - d) \times (1/N) + d \times \left(PR\ (C)/2\right)$$

$$PR\ (B) = (1 - d) \times (1/N) + d \times \left(PR\ (A)/1 + PR(C)/2\right)$$

$$PR\ (C) = (1 - d) \times (1/N) + d \times \left(PR\ (B)/1\right)$$

So

$PR\ (A) = 0.1 + 0.35 \times PR\ (C)$
$PR\ (B) = 0.1 + 0.70 \times PR\ (A) + 0.35 \times PR\ (C)$
$PR\ (C) = 0.1 + 0.70 \times PR\ (B)$

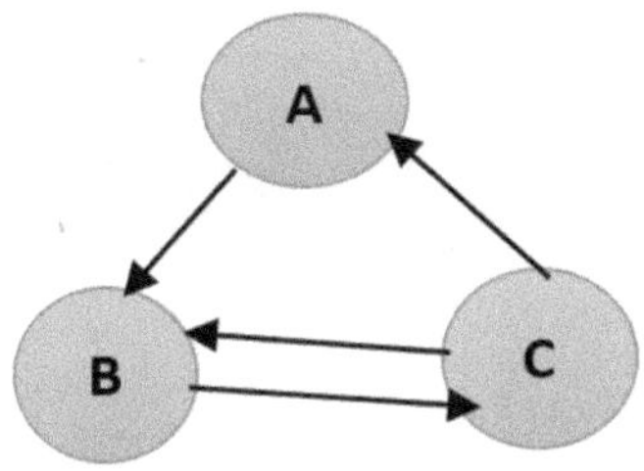

FIGURE 5.10 PageRank example: Graph 2

By solving the above system of linear equations, we get

$$PR\,(A) = 0.2314$$

$$PR\,(B) = 0.3933$$

$$PR\,(C) = 0.3753$$

5.5 PROBLEMS FOR SELF-ASSESSMENT

Q1. Consider the following web graph:

$$V = \{a, b, c\},\ E = \{a \rightarrow b;\ a \rightarrow c;\ b \rightarrow c;\ c \rightarrow b\}$$

Compute PageRank, hub and authority scores for each of the three pages. Also give the relative ordering of the three nodes for each of these scores, indicating any ties.

For PageRank, consider the damping factor as 0.85 and for hubs/authorities, normalize the hub (authority) scores so that the maximum hub (authority) score is 1.

Q2. Compute the hub and authority weights for the following graph:

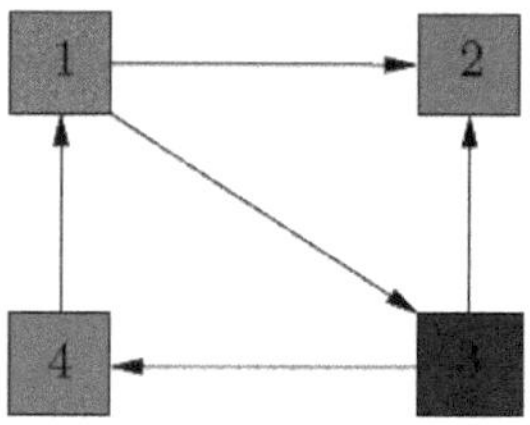

Q3. We have four pages – imaginatively titled Page A, Page B, Page C and Page D. They link to each other as shown in the following diagram. Rank all the pages using PageRank algorithm.

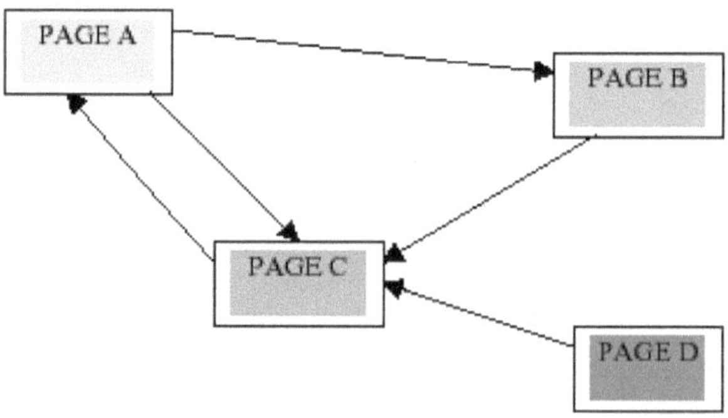

Q4. Suppose that P, Q and R are different web pages. Explain how it can happen that adding a link from P to Q can raise the PageRank of R. Explain how it can happen that adding a link from P to Q can lower the PageRank of R. In both cases, you should show a specific graph where this happens, although you need not work out the actual numerical values.

Q5. Assume $d = 0.85$.

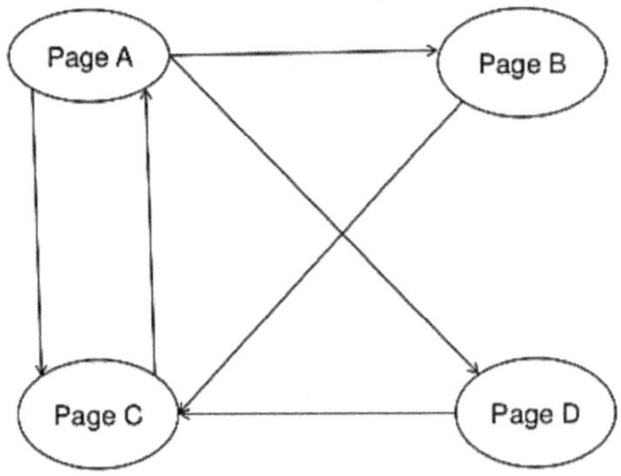

a. Rank all the pages using the PageRank algorithm.
b. Compute the hub and authority weights.

REFERENCES

1. Rezai, D. N. *Hubs and authorities.* https://virgool.io/@deniz.n.rezai/hubs-and-authorities-r6uvlstnwqpe
2. Nithya, T. (2013). Link analysis algorithm for web structure mining. *International Journal of Advanced Research in Computer and Communication Engineering* 2(8), 2950–2954.
3. Manning, C. D., Raghaven, P., & Schutze, H. (2008). Hubs and authorities. *Introduction to Information Retrieval.* Cambridge University Press. https://nlp.stanford.edu/IR-book/html/htmledition/hubs-and-authorities-1.html
4. Lecture #4: HITS algorithm – Hubs and authorities on the internet. Cornell University. http://pi.math.cornell.edu/~mec/Winter2009/RalucaRemus/Lecture4/lecture4.html
5. Wikipedia.PageRank. https://en.wikipedia.org/wiki/PageRank

6 Link Prediction

Social networks are a popular way to model interactions among people in a group or community. They can be visualized as graphs, where a vertex corresponds to a person in some group and an edge represents some form of association between the corresponding persons. Social networks are also very dynamic, because new edges and vertices are added to the graph over time. Understanding the dynamics that drive the evolution of a social network is a complex problem due to a large number of variable parameters.

However, a comparatively easier problem is to understand the association between two specific nodes. For instance, some of the interesting questions that can be posed are: How does the association pattern change over time? What are the factors that drive the associations? How is the association between two nodes affected by other nodes? The problem that we want to tackle here is to **predict the likelihood of a future association between two nodes**, knowing that there is no association between the nodes in the current state of the graph. This problem is commonly known as the **link prediction problem** [1].

6.1 OVERVIEW OF LINK PREDICTION

We are given an unweighted, undirected graph $G = \langle V, E \rangle$ representing the topological structure of a social network in which each edge $e = \langle u, v \rangle \in E$ represents an interaction between u and v that took place at a particular time $t\,(e)$. For two times t and $t' > t$, let $G\left[t, t'\right]$ denote the subgraph of G consisting of all edges with a timestamp between t and t'.

Let t_0, t_0', t_1 and t_1', be four times, where $t_0 < t_0' \le t_1 < t_1$. Then the link prediction task is:

Given network $G[t_0, t_0']$; output a list of edges not present in $G[t_0, t_0']$ that are predicted to appear in the network $G\left[t_1, t_1'\right]$. We refer to $[t_0, t_0']$ as the training interval and $\left[t_1, t_1'\right]$ as the test interval.

6.2 LINK PREDICTION METHODS

We select and present a brief description of eight commonly used prediction methods that use topology information about networks in the prediction process. Throughout this section, the symbols x, y denote nodes, N denotes number of nodes in the network and k is the average degree. $\Gamma(x)$ and $\Gamma(y)$ denote the neighbour sets of these nodes and k_x and k_y denote the degree number of nodes x and y, respectively [2].

6.2.1 Graph Distance

Perhaps the most direct metric for quantifying how close two nodes are is the graph distance. It is defined as the negative of the shortest-path distance from x to y.

$$\text{score}(x,y) = \text{negated}(\text{Length of Shortest Path Between } x \text{ and } y) \tag{6.1}$$

DOI: 10.1201/9781003088066-6

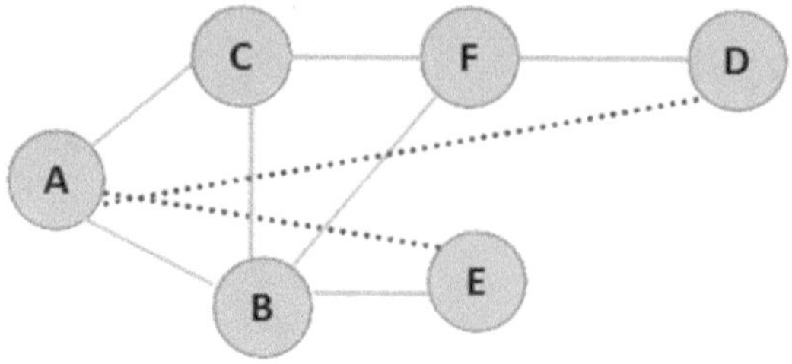

FIGURE 6.1 Graph distance example

Graph distance example:

$$\text{Score}(\text{A},\text{E}) = -2 \, \text{Score}\ (\text{A},\text{D}) = -3$$

As can be seen, the score between node A and E is higher as compared to the score between node A and D; the likelihood of A coming in contact with E is comparatively more.

The measure follows the notion that social networks are ***small worlds****, in which individuals are related through short chains. The use of negated (instead of original) shortest-path distance ensures that the proximity measure $GD(x, y)$ increases as x and y get closer.*

Note that it is inefficient to apply Dijkstra's algorithm to compute the shortest-path distance from *x to y* when *G* has millions of nodes. Instead, we exploit the small-world property of the social network and apply expanded ring search to compute the shortest-path distance from x to y.

Specifically, we initialize $S = \{x\}$ and $D = \{y\}$. In each step, we either expand set S to include its members' neighbours (i.e., $S = S \cup \{v \mid \langle u, v \rangle \in E \wedge u \in S\}$) or expand set D to include its members' inverse neighbours (i.e., $D = D \cup \{u \mid \langle u, v \rangle \in E \wedge v \in D\}$). We stop whenever $S \cap D! = \varnothing$. The number of steps taken so far gives the shortest-path distance. For efficiency, we always expand the smaller set between S and D in each step.

6.2.2 Common Neighbours

The common neighbours predictor captures the notion that two strangers who have a common friend may be introduced by that friend. This introduction has the effect of 'closing a triangle' in the graph and feels like a common mechanism in real life. Newman [3] has computed this quantity in the context of collaboration networks, verifying a positive correlation between the number of common neighbours of x and y at time t, and the probability that x and y will collaborate at some time after t.

$$\text{Score}(x, y) = |\Gamma\ (x) \cap \Gamma\ (y)| \tag{6.2}$$

$\Gamma\ (x)$: Neighbours of x, $\Gamma\ (y)$: Neighbours of y

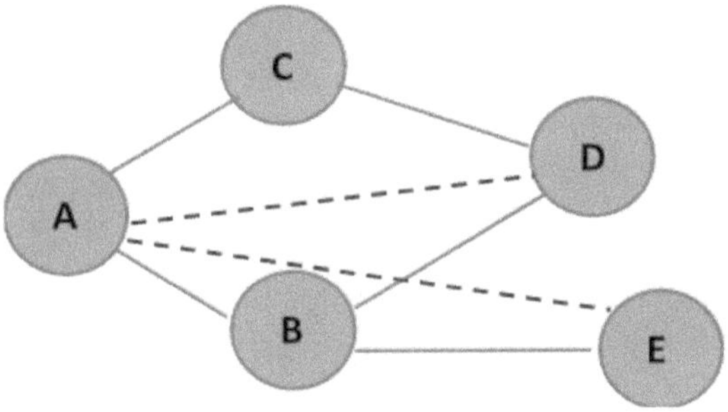

FIGURE 6.2 Example of common neighbours

Common neighbours example: Consider the following network with respect to their scores determined using common neighbours:

$$\text{Score}\ (A,\ D) = |\Gamma\ (A) \cap \Gamma\ (D)| = \{B,\ C\} \cap \{B,\ C\} = 2$$

$$\text{Score}\ (A,\ E) = |\Gamma\ (A) \cap \Gamma\ (E)| = \{B,\ C\} \cap \{B\} = 1$$

As can be seen, the score between node *A* and *D* is higher as compared to the score between node *A* and *E*; the likelihood of *A* coming in contact with *D* is comparatively more.

6.2.3 Jaccard's Coefficient

Jaccard's coefficient – a similarity metric that is commonly used in information retrieval – measures the probability that both *x* and *y* have a feature *f* for a randomly selected feature *f* that either *x* or *y* has. If we take 'features' here to be neighbours, then this measure captures the intuitively appealing notion that the proportion of the co-authors of *x* who have also worked with *y* (and vice versa) is a good measure of the similarity of *x* and *y*.

$$\text{Score}(x,y)\left|\frac{\Gamma\ (\text{x}) \cap \Gamma\ (\text{y})}{\Gamma\ (\text{x}) \cup \Gamma\ (\text{y})}\right| \tag{6.3}$$

where, $\Gamma\ (x) \cup \Gamma\ (y)$ = common friends and $\Gamma\ (x) \cap \Gamma\ (y)$ = total friends

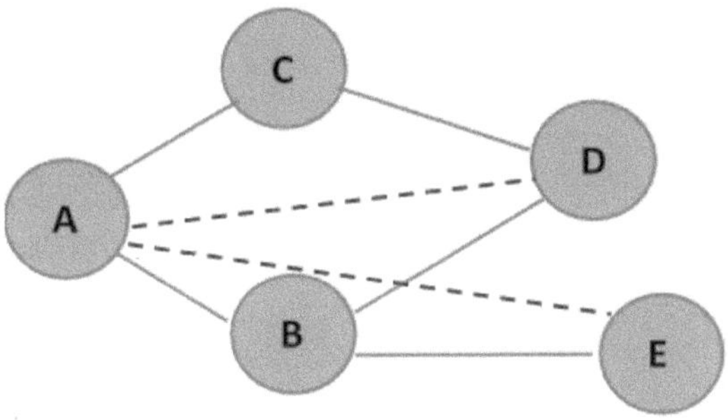

FIGURE 6.3 Example of Jaccard's coefficient

Example: Consider the following network with respect to their scores determined using Jaccard's coefficient:

$$\text{Score}\,(A,\,D) = \frac{|\Gamma\,(A) \cap \Gamma\,(D)|}{|\Gamma\,(A) \cup \Gamma\,(D)|} = \frac{\{B,\,C\} \cap \{B,\,C\}}{\{B,\,C\} \cup \{B,\,C\}} = \frac{2}{2} = 1$$

$$\text{Score}\,(A,\,E) = \frac{|\Gamma\,(A) \cap \Gamma\,(E)|}{|\Gamma\,(A) \cup \Gamma\,(E)|} = \frac{\{B,\,C\} \cap \{B\}}{\{B,\,C\} \cup \{B\}} = \frac{1}{2} = 0.5$$

As can be seen, the score between node A and D is higher as compared to the score between node A and E; the likelihood of A coming in contact with D is comparatively more.

This metric solves the problem where two nodes could have many common neighbours because they have lots of neighbours, not because they are strongly related.

6.2.4 Adamic/Adar (Frequency-Weighted Common Neighbours)

This measure refines the simple counting of common features by weighting rarer features more heavily. The Adamic/Adar predictor formalizes the intuitive notion that rare features are more telling; documents that share the phrase 'for example' are probably less similar than documents that share the phrase 'clustering coefficient'.

If 'triangle closing' is a frequent mechanism by which new edges form in a social network, then for x and y to be introduced by a common friend z, person z will have to choose to introduce the pair $\langle x,\, y\rangle$ from (choose $|\Gamma\,(z)|$ with 2) pairs of his friends; thus, an unpopular person (someone with not a lot of friends) may be more likely to introduce a particular pair of his friends to each other.

$$\text{score}(x,\,y) = \sum_{z \in \Gamma(x) \cap \Gamma(y)} \frac{1}{\log|\Gamma\,(z)|} \tag{6.4}$$

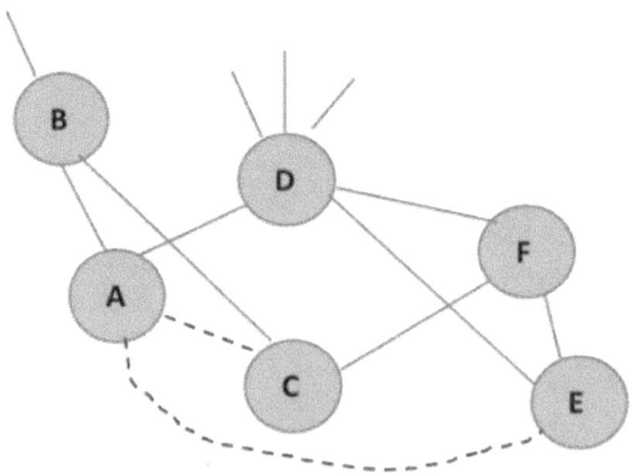

FIGURE 6.4 Example of Adamic/Adar

Example of Adamic/Adar:

$$z = \Gamma(A) \cap \Gamma(C) = B,$$

$$\text{Score}\,(A,\, C) = \frac{1}{\log |\Gamma\,(B)|} = \frac{1}{\log 3} = 2.09$$

$$z = \Gamma(A) \cap \Gamma(E) = D,$$

$$\text{Score}\,(A,\, E) = \frac{1}{\log |\Gamma\,(D)|} = \frac{1}{\log 6} = 1.2$$

6.2.5 Preferential Attachment

One well-known concept in social networks is that users with many friends tend to create more connections in the future. This is due to the fact that in some social networks, like in finance, the rich get richer. We estimate how 'rich' our two vertices are by calculating the multiplication between the number of friends ($|\Gamma(x)|$) or followers that each vertex has. It may be noted that the similarity index does not require any node neighbour information; therefore, this similarity index has the lowest computational complexity.

$$\text{Score}(x, y) = |\Gamma\,(x)| * |\Gamma(y)| \tag{6.5}$$

$AB < AC$, the link between A and C is more probable than the link between A and B because C has many more neighbours than B.

6.2.6 Katz (Exponentially Damped Path Counts)

This heuristic defines a measure that directly sums over collection of paths, exponentially damped by length to count short paths more heavily. The Katz measure is a variant of the shortest-path measure. The idea behind the Katz measure is that

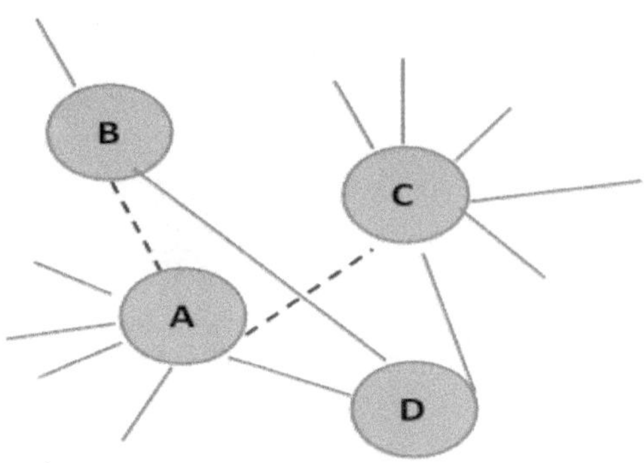

FIGURE 6.5 Example of preferential attachment

the more paths there are between two vertices and the shorter these paths are, the stronger the connection.

This method takes lengths of all paths between each pair of nodes into consideration. According to equation (6.6), the number of paths between node x and node y with length l (written as | paths$\langle l \rangle$ xy |) is calculated and then multiplied by a factor. By summing up all the results for the given two nodes with path length from 1 to ∞, a prediction score for the pair of nodes (x, y) is obtained. Katz is a prediction method based on the topology of a whole network and thus its calculation is more complex than other methods in this chapter. The complexity is mainly determined by the matrix inversion operator, which is $(N3)$:

$$\text{Score}(x, y) = \sum_{l=1}^{\infty} \beta^l . \left|\text{paths}_{xy}^l\right| \tag{6.6}$$

A very small β *yields predictions much like common neighbours, because paths of length three or more contribute very little to the summation.*

Example of the Katz measure:

$$\text{No. of hops} = 2$$

$$\text{Damping factor} = 1/2$$

$$\text{path}_{A,D}^2 = 2,\ \text{path}_{A,D}^3 = 2,$$

$$S = \frac{1}{2} . 2 + \frac{1}{4} . 2 + \ldots$$

$$\text{path}_{A,E}^2 = 1,\ \text{path}_{A,E}^3 = 1,$$

$$\text{S} = \frac{1}{2} . 1 + \frac{1}{4} . 1 + \ldots$$

The parameter β, as shown in equation (6.6) is used to adjust the weight of path with different length. When an extremely small β is chosen, the longer paths will contribute less to the score in comparison to shorter ones so that the result will be

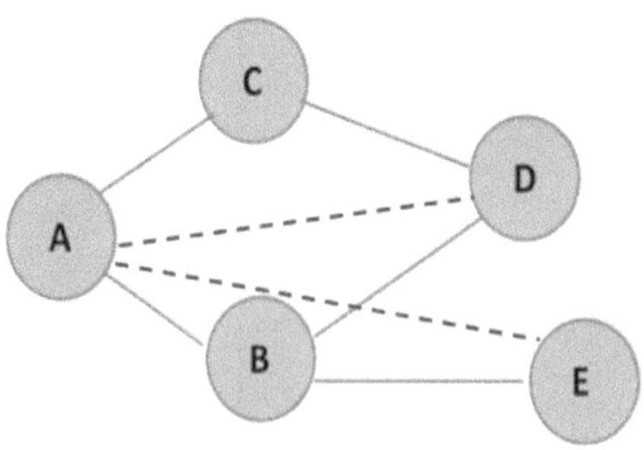

FIGURE 6.6 Example of the Katz measure

close to the common neighbours. It is one of the prediction methods that, as it will be shown in further sections, achieves high prediction accuracy in many experiments.

6.2.7 Hitting Time

A random walk starts at a node x and iteratively moves to a neighbour of x chosen uniformly at random. The hitting time $H_{x,y}$ from x to y is the expected number of steps required for a random walk starting at x to reach y.

$$\text{Score}(x, y) = -\text{H}_{xy} \tag{6.7}$$

$-\text{H}_{xy}$ is expected time/step for random walk from x to reach y.

(Symmetric version) The commute time $C_{x,y}$ from x to y is the expected number of steps to travel from x to y and from y to x.

$$\text{Score }(x, y) = -\left(\text{H}_{x,y} + \text{H}_{y,x}\right) \tag{6.7.1}$$

One difficulty with hitting time as a measure of proximity is that H_{xy} is quite small whenever y is a node with a large stationary probability π_y, regardless of the identity of x (i.e., for a node y at which the random walk spends a considerable amount of time in the limit, the random walk will soon arrive at y, almost no matter where it starts. Thus, the predictions made based upon H_{xy} tend to include only a few distinct nodes y). To counterbalance this phenomenon, we also consider normalized versions of the hitting and commute times, by defining

$$\text{score}(x, y) = -\text{H}_{xy} \cdot \pi_y \tag{6.7.2}$$

We can also consider stationary-normed versions

$$\text{score}(x, y) = -\text{H}_{xy} \cdot \pi_y \tag{6.7.3}$$

$$\text{score}(x, y) = -\left(\text{H}_{xy} \cdot \pi_y + \text{H}_{yx} \cdot \pi_x\right)$$

The hitting time and commute time measures are sensitive to parts of the graph far away from x and y → periodically ***reset the walk***

Random walk on G_{collab} that starts at x and has a probability of α of returning to x at each step.

6.2.8 Rooted (Personalized) PageRank

Another difficulty with using the measures based on hitting time and commute time is their sensitive dependence to parts of the graph far away from x and y, even when x and y are connected by very short paths. A way of counteracting this difficulty is to allow the random walk from x to y to periodically 'reset', returning to x with a fixed probability α at each step; in this way, distant parts of the graph will almost never be explored.

Starts from x, with probability $(1 - a)$ moves to a random neighbour and with probability a returns to x

$$\text{Score}(x, y) = \text{stationary probability of } y \text{ in a rooted PageRank} \tag{6.8}$$

Random resets form the basis of the PageRank measure for web pages, and we can adapt it for link prediction. Similar approaches have been considered for personalized PageRank, in which one wishes to rank web pages based both on overall 'importance' (the core of PageRank) and relevance to a particular topic or individual, by biasing the random resets towards topically relevant or bookmarked pages

6.3 OTHER METRICS

All of the methods presented in this section follow a similar approach. The required input for each method is the adjacency matrix that represents a network in which there are only 0 and 1 (0 when there is no link between two given nodes and 1 when the links between two given nodes exist). The output of each method is a similarity matrix in which each element represents the similarity score of a pair of nodes within the network and it is calculated according to the equation used in a given method.

6.3.1 Friends Measure

When looking at two vertices in a social network, we can assume that the more connections their neighbourhoods have with each other, the higher the chances the two vertices are connected. We take the logic of this statement and define the friends measure as the number of connections between u and v neighbourhoods. One can notice that in undirected networks, the friends measure is a private case of the Katz measure where $\beta = 1$ and $l_{\max} = 2$.

$$\text{friends measure}(u,v) = \sum_{x \in \Gamma(u)} \sum_{y \in \Gamma(v)} \delta(x,y) \tag{6.9}$$

When we define the function $\delta(x,y)$ as:

$$\delta(x,y) = \begin{cases} 1, \textit{ if } x = y \text{ or } (x,y) \in E \text{ or } (y,x) \in E \\ 0, \text{ otherwise} \end{cases} \tag{6.10}$$

6.3.2 Cosine Similarity

The idea of this method is based on the dot product of two vectors. It is often used to compare documents in text mining. In a network prediction problem, this method is expressed as

$$\text{Score}(\text{x, y}) = \frac{|\Gamma(\text{x})| \cdot |\Gamma(\text{y})|}{\|\Gamma(\text{x})\| \times \|\Gamma(\text{y})\|} = \frac{|\Gamma(\text{x})| \cap |\Gamma(\text{y})|}{\sqrt{|\Gamma(\text{x})||\Gamma(\text{y})|}} \tag{6.11}$$

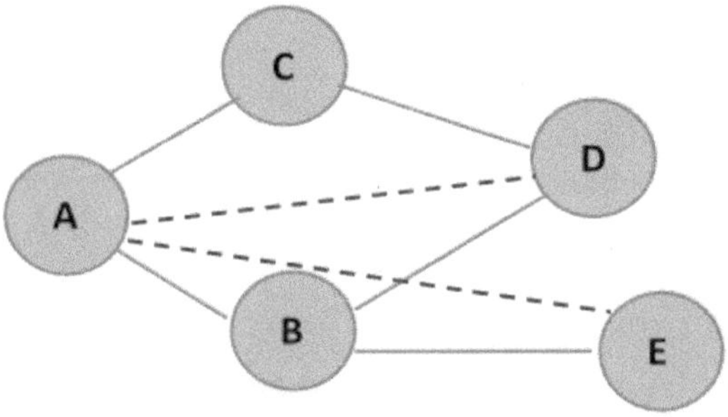

FIGURE 6.7 Example of cosine similarity

For each pair of nodes with common neighbours, this method will perform a vector multiplication and thus the complexity is $\left(Nk^3\right)$.
Example of cosine similarity:

$$\text{Score}(A,D)=\frac{|\Gamma\ (\text{A})|\cdot|\Gamma\ (D)|}{\|\Gamma\ (\text{A})\|\times\|\Gamma\ (D)\|}=\frac{|(C,B)|\cap|\ (C,B)|}{\|(\text{C},\text{B})\|\times\|(C,B)\|}=\frac{2}{\sqrt{2\times 2}}=\frac{2}{2}=1$$

$$\text{Score}(A,E)=\frac{|\Gamma\ (\text{A})|\cdot|\Gamma\ (E)|}{\|\Gamma\ (\text{A})\|\times\|\Gamma\ (E)\|}=\frac{|(C,B)|\cap|(B)|}{\|(\text{C},\text{B})\|\times\|(B)\|}=\frac{1}{\sqrt{2\times 1}}=\frac{1}{\sqrt{2}}=0.71$$

As can be seen, the score between node *A* and *D* is higher as compared to the score between node *A* and *E*; the likelihood of *A* coming in contact with *D* is comparatively more.

6.3.3 Sørensen Index

This index is designed for comparing the similarity of two samples and originally used in analysis of plant sociology. The complexity of this method is $\left(Nk^2\right)$. It is defined as

$$\text{Score}(x,y)=\frac{2|\Gamma\ (\text{x})\cap\Gamma\ (\text{y})|}{k_x+k_y}\tag{6.12}$$

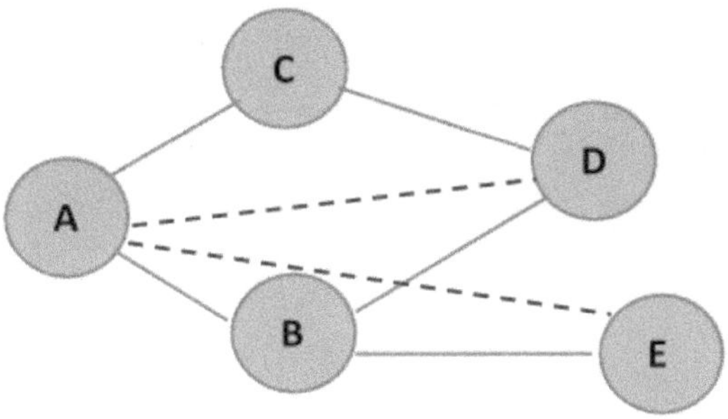

FIGURE 6.8 Example of the Sørensen Index

Example of the Sørensen Index:

$$\text{Score}(A,D)=\frac{2*2}{2+2}=\frac{4}{4}=1$$

$$\text{Score}(A,E)=\frac{2*1}{2+1}=\frac{2}{3}=0.66$$

As can be seen, the score between node A and D is higher as compared to the score between node A and E; the likelihood of A coming in contact with D is comparatively more.

6.3.4 Hub Promoted Index

The Hub Promoted Index is proposed for analyzing metabolic networks. The property of this index is that the links adjacent to hubs are likely to obtain a higher similarity score. The complexity of the method is (Nk^2). It is expressed as

$$\text{Score}(x,y)=\frac{|\Gamma(x)\cap\Gamma(y)|}{\min\{k_x,\ k_y\}} \tag{6.13}$$

Example of the Hub Promoted Index:

$$\text{Score}(A,D)=\frac{2}{2}=1$$

$$\text{Score}(A,E)=\frac{1}{1}=1$$

As can be seen, the score between node A and D and the score between node A and E is the same; the likelihood of A coming in contact with D and A coming in contact with E is the same.

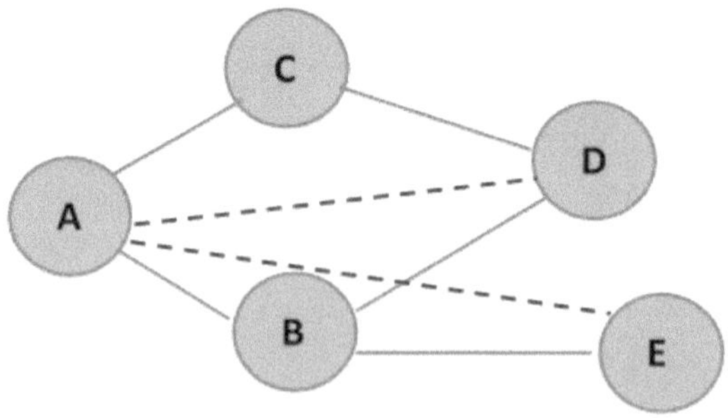

FIGURE 6.9 Example of the Hub Promoted Index

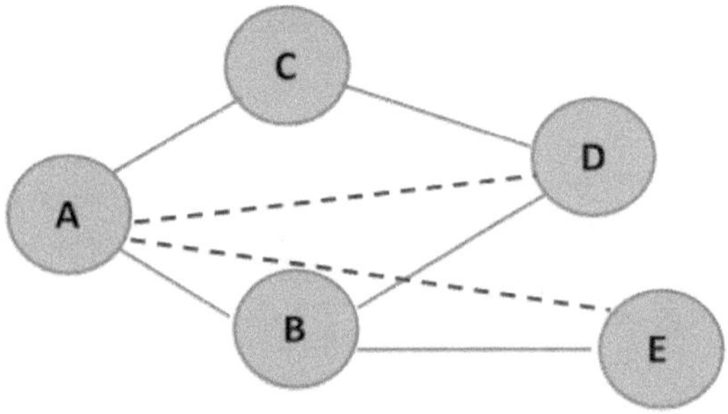

FIGURE 6.10 Example of the Hub Depressed Index

6.3.5 Hub Depressed Index

The Hub Depressed Index is an approach that uses the idea of a hub in a totally different manner than the Hub Promoted Index. It gives links adjacent to the hub a lower score. Its complexity is the same as the Hub Promoted Index, ($Nk2$). It is defined as

$$\text{Score}(x,y) = \frac{|\Gamma\,(\text{x}) \cap \Gamma\,(\text{y})|}{\max\{k_x,\, k_y\}} \tag{6.14}$$

Example of the Hub Depressed Index:

$$\text{Score}(A,D) = \frac{2}{2} = 1$$

$$\text{Score}(A,E) = \frac{1}{2} = 0.5$$

As can be seen, the score between node A and D is higher as compared to the score between node A and E; the likelihood of A coming in contact with D is comparatively more.

6.3.6 Leicht–Holme–Newman Index

The Leicht–Holme–Newman Index [4] was proposed to quantify the similarity of nodes in networks. It is based on the concept that two nodes are similar if their immediate neighbours in the network are themselves similar. As another common neighbour-based method, its complexity is ($Nk2$). It is defined as

$$\text{Score}(x,y) = \frac{|\Gamma\,(\text{x}) \cap \Gamma\,(\text{y})|}{k_x * k_y} \tag{6.15}$$

Example of the Leicht–Holme–Newman Index:

$$\text{Score}(A,D) = \frac{2}{2*2} = \frac{2}{4} = 0.5$$

$$\text{Score}(A,E) = \frac{1}{2*1} = \frac{1}{2} = 0.5$$

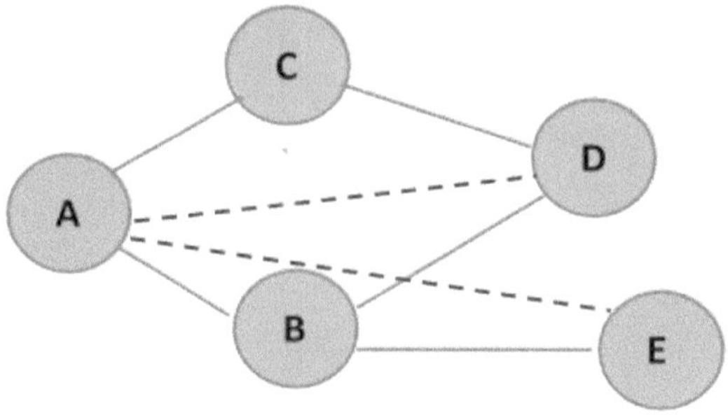

FIGURE 6.11 Example of the Leicht–Holme–Newman Index

As can be seen, the score between node A and D and the score between node A and E is the same; the likelihood of A coming in contact with D and A coming in contact with E is the same.

6.4 PREDICTION PERFORMANCE METRICS

In order to measure the performance of a prediction method, we need to use historical network data. Link prediction is a time-related activity; therefore, we should use a time-stamped data set, and according to the time stamp, separate the data into two sets: $G_{t,t_1}\ (V, L_1)$ as a training set for prediction methods and $G_{t_1,t_2}\ (V, L_2)$ as an unknown future network for testing where $t < t_2$. Those two networks must consist of the same set of nodes V.

The number of possible links that is denoted by U is $|V| * (|V| - 1)/2$

The link prediction method, in principle, provides a similarity score for each non-existing link $(U - L_1)$ and for most methods; a higher score means a higher likelihood that the link will appear in the future. Final prediction is done by ordering this score list and selecting top N links with the highest score. *AUC* (area under the receiver operating characteristic curve) is used for quantifying the accuracy of a prediction method.

In the context of network link prediction, *AUC* can be interpreted as the probability that a randomly chosen missing link $(L_1 \cup L_2 - L_1)$ is given a higher similarity score than a randomly chosen pair of unconnected links $(\mathrm{U} - (L_1 \cup L_2))$.

AUC is calculated as

$$AUC = \frac{n' + 0.5n''}{n} \tag{6.16}$$

Where n is the number of times that we randomly pick a pair of links from a missing links set and unconnected links set; n' is the number of times that the missing link got a higher score than an unconnected link, while n'' is the number of times when they are equal.

The *AUC* value will be 0.5 if the score is generated from an independent and identical distribution. Thus, the degree to which the *AUC* exceeds 0.5 indicates how much better the predictions are when compared to prediction by chance.

6.5 PROBLEMS FOR SELF-ASSESSMENT

Q1. Compute score (u, v) for all edges in the graph below. Remove the edges with a low score. And also predict the links between disconnected nodes in the future.

a. Calculate Katz score for node 'L'.

b. Calculate Adamic/Adar score for node 'D'.

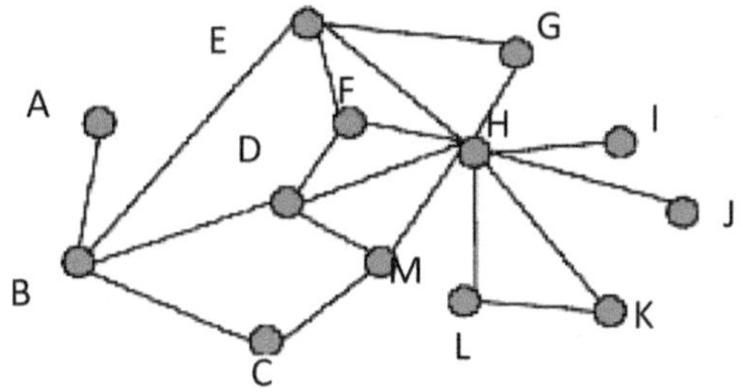

Q2. Consider the graph below and justify your answer for the following questions:

a. What will be the score of node D and node E using Adamic/Adar of time stamp t_{i+1}?

b. Predict any two-probable links in the graph of time stamp t_{i+1} using Jaccard's coefficient.

c. Using Katz calculates score, explain which will be the most probable link to connect G and why.

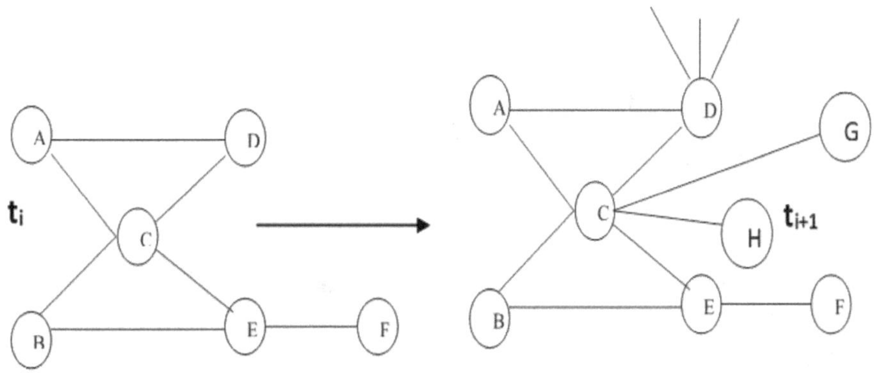

REFERENCES

1. Link prediction algorithms. (2014). http://be.amazd.com/link-prediction/
2. Gao, F., Musial, K., Cooper, C., & Tsoka, S. (2015). Link prediction methods and their accuracy for different social networks and network metrics. *Scientific Programming*. https://doi.org/10.1155/2015/172879
3. Newman, M. E. (2003). The structure and function of complex networks. *SIAM Review*, 45(2), 167–256.
4. Leicht, E. A., Holme, P., & Newman, M. E. (2006). Vertex similarity in networks. *Physical Review E*, 73(2), 026120.

7 Community Detection

7.1 OVERVIEW OF COMMUNITY

Community is formed by individuals such that those within a group interact with each other more frequently than with those outside the group, also known as group, cluster, cohesive subgroup and module in different contexts [1].

A community, with respect to graphs, can be defined as a subset of nodes that are densely connected to each other and loosely connected to the nodes in the other communities in the same graph.

In the study of complex networks, a network is said to have *community structure* if the nodes of the network can be easily grouped into (potentially overlapping) sets of nodes such that each set of nodes is densely connected internally. In the particular case of a *non-overlapping* community finding, this implies that the network divides naturally into groups of nodes with dense connections internally and sparser connections between groups. But *overlapping* communities are also allowed. The more general definition is based on the principle that pairs of nodes are more likely to be connected if they are both members of the same community(ies), and less likely to be connected if they do not share communities. A related but different problem is community search, where the goal is to find a community that a certain vertex belongs to [2].

Qualitatively, a community is defined as a subset of nodes within the graph such that connections between the nodes are denser than connections with the rest of the network. The detection of the community structure in a network is generally intended as a procedure for mapping the network [3] into a tree (Figure 7.1). In this tree (called a dendrogram in the social sciences), the leaves are the nodes whereas the branches join nodes or (at higher level) groups of nodes, thus identifying a hierarchical structure of communities nested within each other.

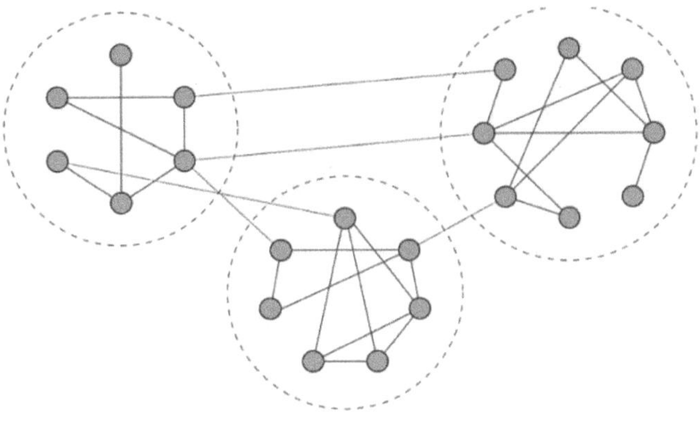

FIGURE 7.1 Detection of a community

DOI: 10.1201/9781003088066-7

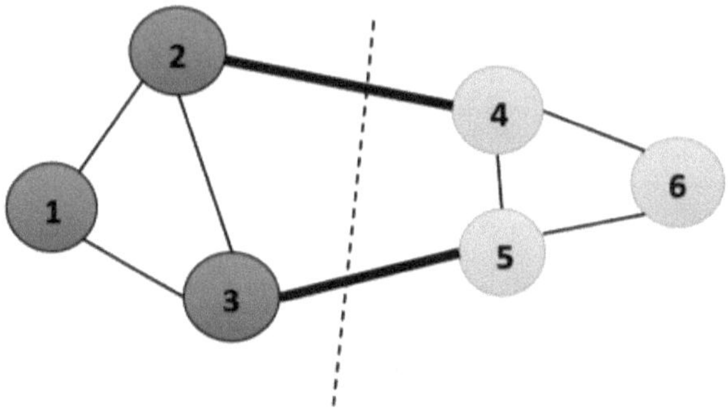

FIGURE 7.2 Community detection

Why are there communities in social media?: Communities are also formed on social media platforms because human beings are social. Communities are easy to use because social media allows people to extend their social life in unprecedented ways. As we know, it is difficult to meet friends in the physical world, but much easier to find friends online with similar interests on social network platforms. Interactions between nodes can help determine communities.

Type of communities in social media: There are two types of groups in social media: explicit groups, formed by user subscriptions and implicit groups, implicitly formed by social interactions. Some social media sites allow people to join groups, so it is necessary to extract groups based on network topology because not all sites provide a community platform, not all people want to make an effort to join groups and groups can change dynamically.

But network interaction provides rich information about the relationship between users. It can:

- Complement other kinds of information.
- Help network visualization and navigation.
- Provide basic information for other tasks.

Community detection: This is discovering groups in a network where individuals' group memberships are not explicitly given. Detecting communities in a network is one of the most important tasks in network analysis. In a large-scale network [4], such as an online social network, we could have millions of nodes and edges. Detecting communities in such networks becomes a herculean task. Therefore, we need community detection algorithms that can partition the network into multiple communities.

7.2 TAXONOMY OF COMMUNITY CRITERIA

Criteria vary depending on the tasks. Roughly, community detection methods can be divided into four categories (not exclusive): node-centric, group-centric, network-centric and hierarchy-centric community detection.

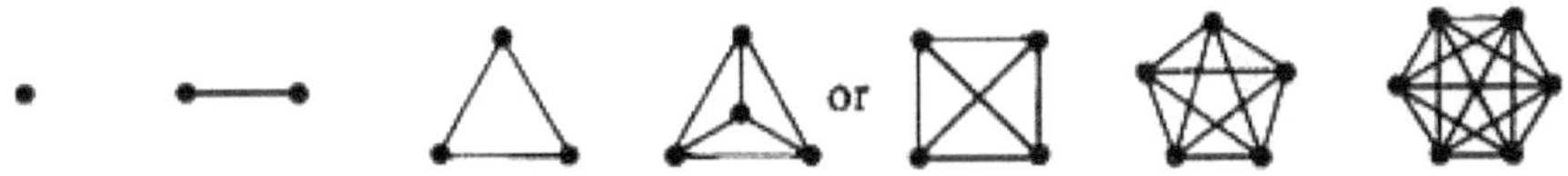

FIGURE 7.3 Cliques

7.2.1 Node-Centric Community Detection

If each node in a group satisfies certain properties, they form a community. This is commonly used in traditional social network analysis (SNA). Node-centric communities can be detected by the following methods [5].

Clique

To understand a clique, we need to first learn about the subgraph. Given an undirected graph $G = (V, E)$ where V denotes the set of vertices and E the set of edges, the graph $G = (V_1, E_1)$ is called a subgraph of G if V_1 is a subset of V, E_1 is a subset of E and for every edge (v_i, v_j) from E_1 the vertices v_i, v_j belong to G_1. A subgraph G_1 is said to be complete if there is an edge for each pair of vertices. A complete subgraph is also called a clique.

Complete Mutuality

In this, we find cliques as communities. A clique is a narrow exclusive circle or group of persons, especially one held together by common interests, views or purposes, such as high school cliques. In graph language, a clique is a maximum complete subgraph in which all nodes are adjacent to each other.

In this example (Figure 7.4) Nodes 5, 6, 7 and 8 form a clique. Finding the maximum clique in a network is *NP* hard. Straightforward implementation to find cliques is very expensive in time complexity.

Maximum Clique (Algorithm)

In a clique of size k, each node maintains degree >= $k - 1$. Nodes with degree < k −1 will not be included in the maximum clique. Recursively apply the following pruning procedure:

1. Sample a subnetwork from the given network, and find a clique in the subnetwork, say, by a greedy approach

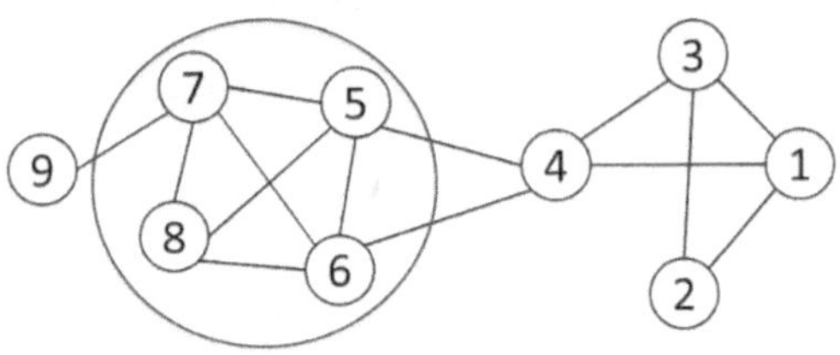

FIGURE 7.4 Clique recognition

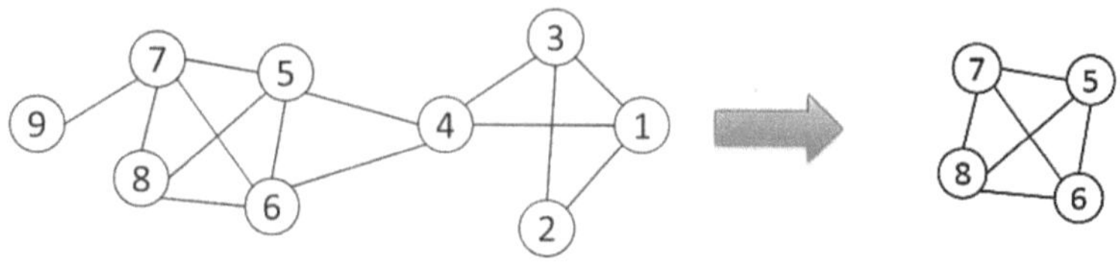

FIGURE 7.5 Maximum clique

2. Suppose the clique above is size k, in order to find out a larger clique, all nodes with degree <= k −1 should be removed.
3. Repeat until the network is small enough.

Maximum clique example:

Suppose we sample a subnetwork with nodes $\{1\ldots5\}$ and find a clique $\{1, 2, 3\}$ of size 3. In order to find a clique > 3, remove all nodes with degree $<= 3-1=2$

- Remove nodes 2 and 9
- Remove nodes 1 and 3
- Remove node 4

Many nodes will be pruned (dead) as social media networks follow a power law distribution for node degrees.

Clique Percolation Method

A clique, in a very strict definition, is unstable. We normally use cliques as a core or a seed to find larger communities. The clique percolation method (CPM) is such a method to find overlapping communities.

CPM ALGORITHM: EACH CONNECTED COMPONENT IN THE CLIQUE GRAPH FORMS A COMMUNITY.

Input: A parameter k, and a network

Procedure:

- Find out all cliques of size k in a given network.
- Construct a clique graph. Two cliques are adjacent if they share k −1 nodes.

CPM example:

The clique structure, where there must be an edge for each pair of vertices, shows many restrictions in real-life modelling. So, alternative approaches were suggested in order to relax the clique concept, such as k – clique, k – clan/k – club and k – plex.

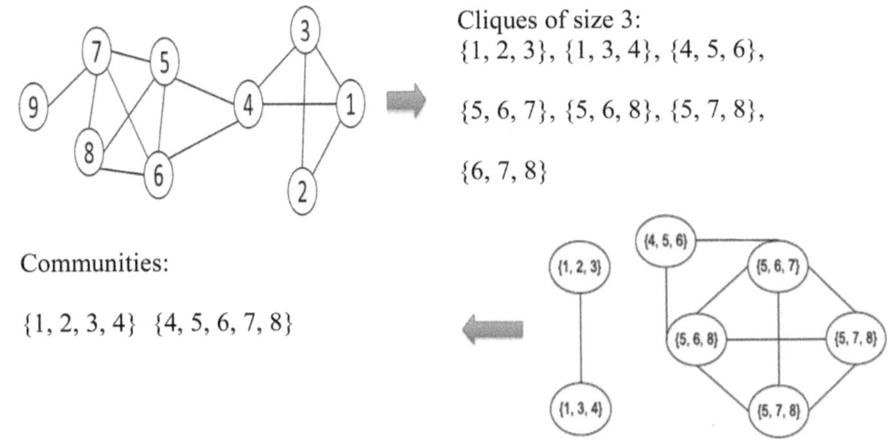

FIGURE 7.6 CPM example

Reachability of Members

Any node in a group should be reachable in k hops. It can be measured by k-clique, k-club and k-clan.

***k*-clique:** a maximal subgraph in which the largest geodesic distance between any nodes <= k. A k-clique is a subset of vertices C such that, for every i, j belonging to C, the distance $d\left(i, j\right) <= k$. The *1*-clique is identical to a clique, because the distance between the vertices is one edge. The 2-clique is the maximal complete subgraph with a path length of one or two edges. The path distance two can be exemplified by the 'friend of a friend' connection in social relationships.

In social websites like LinkedIn, each member can reach their own connections and also connections two and three degrees away. The increase of the value k corresponds to a gradual relaxation of the criterion of clique membership.

A limitation of the k-clique model is that some vertices may be distant from the group, i.e., the distance, between two nodes, may correspond to a path involving nodes that do not belong to the k-clique. To overcome this handicap, diameter- based models called k-club and k-clan were introduced.

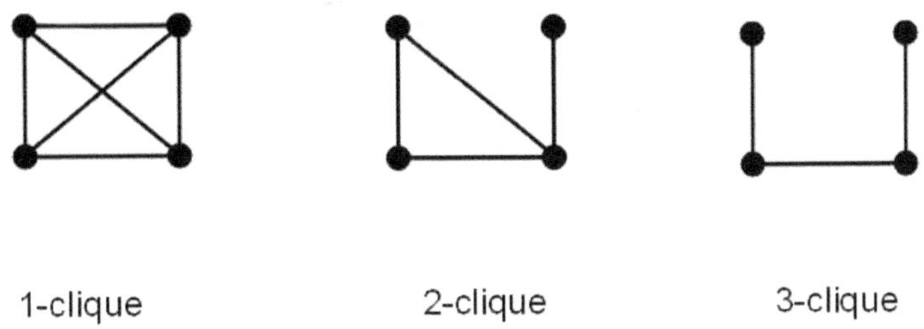

FIGURE 7.7 Example of k-cliques

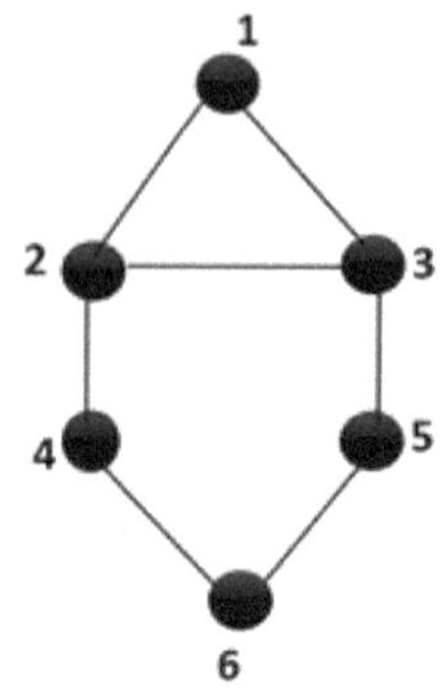

FIGURE 7.8 Example graph for *k*-clan/*k*-club

***k*-clan/*k*-club:** A substructure of *diameter* <= k.
Let's first learn a few definitions:

$d\ (u,v)$: The length of the shortest path between vertices u and v in G
diam (G): The diameter of G; max $d(u,\ v)$

To find all *k*-clan/*k*-club, first all the *k*-cliques S_i must be found and then the restriction diameter $(G[S]) <= k$ applied to remove the undesired *k*-cliques.

Another approach to the diameter models is the *k*-club that is defined as a subset of vertices S such that diameter $(G[S])$ <= k. In Figure 7.8, two 2-cliques: {1,2,3,4,5} and {2,3,4,5,6}, one 2-clan: {2,3,4,5,6} and three 2-clubs: {1,2,3,4}, {1,2,3,5} and {2,3,4,5,6} can be found. In Figure 7.8, from the 2-clan, the 2-clique {1, 2, 3, 4, 5} was removed because d (4, 5) = 3.

Example of reachability:

Cliques: $\{1, 2, 3\}$
2-cliques: $\{1, 2, 3, 4, 5\}$, $\{2, 3, 4, 5, 6\}$
2-clubs: $\{1, 2, 3, 4\}$, $\{1, 2, 3, 5\}$, $\{2, 3, 4, 5, 6\}$

A *k*-clique might have diameter larger than *k* in the subgraph commonly used in traditional SNA. Note that the path of length *k* or less linking a member of the *k*-clique to another member may pass through an intermediary who is not in the group (e.g., for nodes 4 and 5).

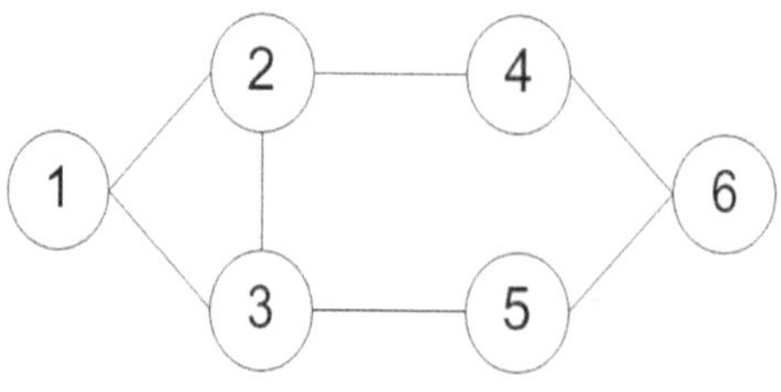

FIGURE 7.9 Example graph for reachability

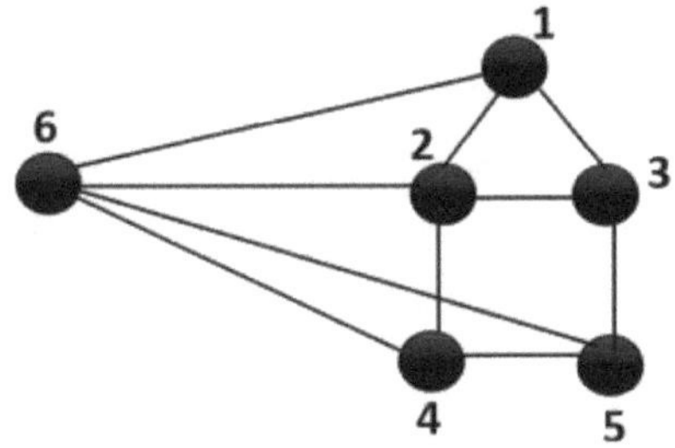

FIGURE 7.10 Example for nodal degree

Nodal Degrees

Community can also be detected by nodal degrees such as k-plex, k-core. Each node should have a certain number of connections to nodes within the group. k-core is a substructure that each node connects to at least k members within the group. k-plex can be detected for a group with n_s nodes, each node should be adjacent no fewer than $n_s - k$ in the group.

k-plex: An alternative way of relaxing a clique is the k-plex concept that takes into account the vertices degree. The degree of a vertex of a graph is the number of edges incident on the vertex, and is denoted by $\deg(v)$. The maximum degree of a graph G is the maximum degree of its vertices and is denoted by Δ(G). Similarly, minimum degree is the minimum degree of its vertices and is denoted by δ(G).

A subset of vertices S is said to be a k-plex if the minimum degree in the induced subgraph >= $|S| - k$.

Example: In Figure 7.10, the graph has 6 vertices, $|S| = 6$, and the degree of vertices 1, 3, 4 and 5 does not exceed the value 3. So, the minimum degree in the induced subgraph is 3. Thus, for $|S| = 6$, $k = 3$ is obtained.

***k*-core:** given a graph G and an integer k, k-cores of the graph are connected components that are left after all vertices of degrees less than k have been removed. A k-core is a $(n_s - k)$ plex.

Example *k*-core: all the nodes less than degree 3 have been removed to get the 3-core graph.

Relative frequency of within-outside ties: LS sets, Lambda sets

Luccio and Sami introduced [6] the concept of an LS set, which is a subgraph $C = (V, E)$ such that $k_v^{\text{int}(C)} > k_v^{\text{ext}(C)}$; $\forall v \in V$.

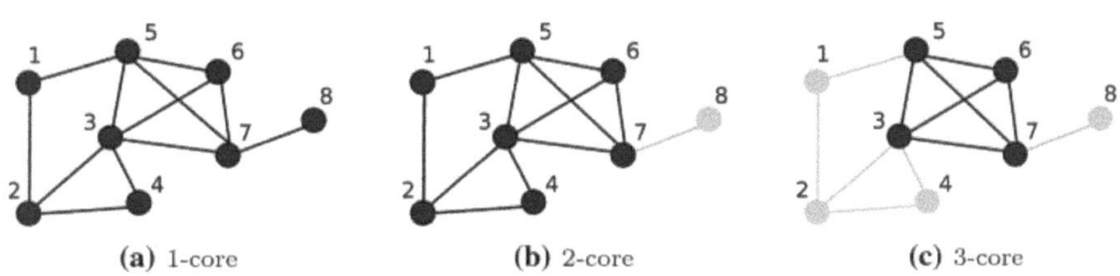

FIGURE 7.11 k-core: example 1

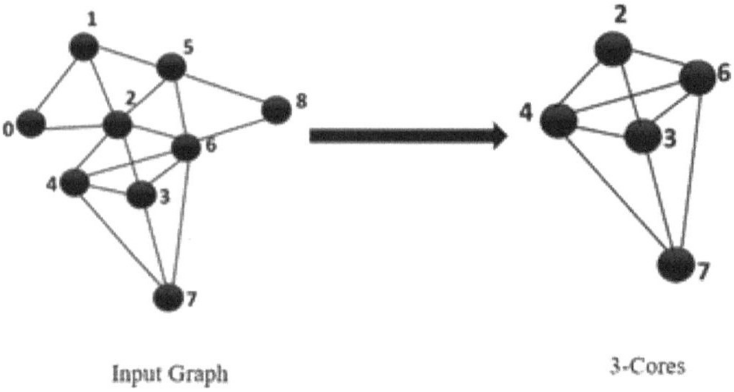

FIGURE 7.12 *k*-core: example 2

where the internal degree $k_v^{\text{int}(C)}$ of a vertex v in a community C is the number of edges connecting v to other vertices of C. And the external degree $k_v^{\text{ext}(C)}$ of a vertex v in a community C is the number of edges connecting v to the rest of the graph.

A vertex $v \epsilon C$ where $k_v^{\text{ext}(C)} = 0$; has thus found a good community in C. If on the other hand, $k_v^{\text{int}(C)} = 0$; v is disjoint from C and should be assigned to another community. This is a strong community [7] but they also define a weak community C such that $k_{\text{int}}^C > k_{\text{ext}}^C$. The edge connectivity of two vertices to be the minimum number of edges one has to remove in order to disconnect them [8].

LS sets: let $G\ (V,\ E)$ be a graph.

The essence of the idea is that an *LS* set [8] may be thought of as the union of its subsets, and this union is 'better' than any subset because it has fewer connections to the outside. Note that the set of all nodes V is an *LS* set, and that every singleton subset $\{u\}$, $u \in V$, is also an *LS* set. We refer to these as 'trivial' *LS* sets. Connected graphs have $(I/1 + 1)$ trivial *LS* sets. As a network model of a cohesive subset, an *LS* set has many appealing characteristics. One important property is that each member of an *LS* set is required to have more connections with other members than with non-members. For example, the graph in Figure 7.13 contains the following 16 LS sets:

$$\{1\}\ \{2\}\ \{3\}\ \{4\}\ \{5\}\ \{6\}\ \{7\}\ \{8\}\ \{9\}\ \{10\}\ \{11\}\ \{12\}$$

$$\{1\ 2\ 3\ 4\}$$

$$\{1\ 2\ 3\ 4\ 5\ 6\ 7\ 8\}\ \{9\ 10\ 11\ 12\}$$

$$\{1\ 2\ 3\ 4\ 5\ 6\ 7\ 8\ 9\ 10\ 11\ 12\}.$$

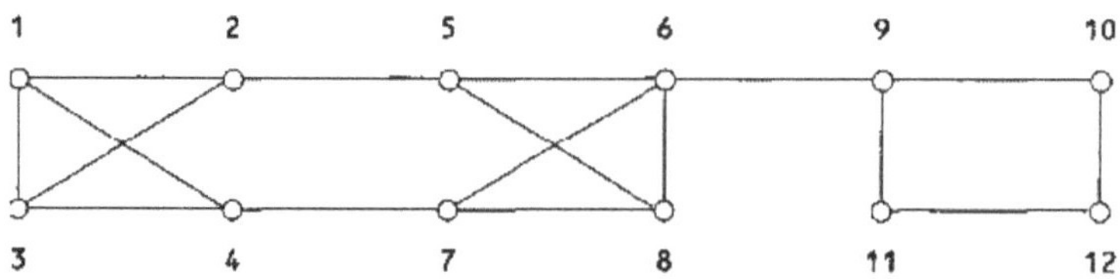

FIGURE 7.13 Example graph to form LS sets and Lambda sets

LS sets: ties concept – more inside than outside. A set of nodes S in a social network is an LS set if each of its proper subsets has more ties to its complement within S than to outside of S.

A Lambda set is a subgraph such that any two vertices in the subgraph have greater edge connectivity than any pair of vertices where one belongs to the subgraph and the other not. The weakness of this definition is that two vertices in a Lambda set are not necessarily adjacent to each other [6]. A relaxed definition is Lambda sets that require the computation of edge-connectivity between any pair of nodes via minimum-cut, maximum-flow algorithm. Comparing within to outside subgroup ties cohesive subgroups should be relatively cohesive within compared to outside.

Lambda set: connectivity concept: should be hard to disconnect by removal of edges.

Lambda (i, j): the number of edges that must be removed from the graph in order to leave no path between any two nodes. Lambda (i, j) > Lambda (k, l) where i, j, k belong to the subgroup and k does not.

7.2.2 Group-Centric Community Detection

Dense communities: cliques, clubs and clans are examples of connected dense. We focus on subgraphs that should be disconnected. We can utilize the brute-force clique identification algorithm

$$\text{Density } \gamma = \frac{|E|}{\binom{|V|}{2}}$$

Consider the connections within a group as a whole. The group has to satisfy certain properties without zooming into node-level. The group-centric criterion requires the whole group to satisfy a certain condition (e.g., the group density >= a given threshold). A subgraph Gs (Vs, Es) is a $\gamma -$ dense $\geq \gamma$ quasi-clique if

$$\frac{|\text{Es}|}{|\text{Vs}|\left(|\text{Vs}| - 1\right)/2} \geq \gamma$$

Consider the connections within a group as whole; some nodes may have low connectivity. A subgraph with V_s nodes and E_s edges is a γ-dense quasi-clique [9] if

Recursive pruning:

Sample a subgraph, find a maximal $\gamma -$ dense quasi $-$ clique
the resultant size $= k$
Remove the nodes that
whose degree $< k\gamma$
all their neighbours with degree $< k\gamma$

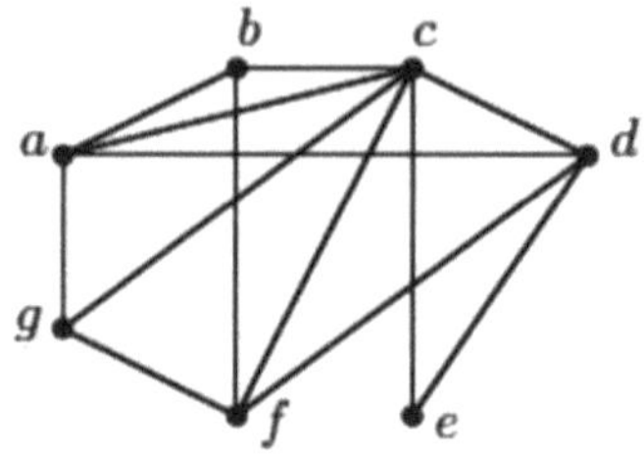

FIGURE 7.14 Example graph G

For example: for graph G, quasi-clique with $\gamma = 0:6$ and min-size $k = 5$, vertices a; b; c; d; f; g form a maximal quasi-clique.

7.2.3 Network-Centric Community Detection

Network-centric criterion needs to consider the connections within a network globally. The goal of network-centric community detection is to partition the whole network into several disjoint sets. There are various approaches through which we can do network-centric community detection:

1. Clustering based on vertex similarity
2. Latent-space models
3. Block model approximation
4. Spectral clustering
5. Modularity maximization

Clustering Based on Vertex Similarity

Clustering based on vertex similarity can be done by applying k-means or similarity-based clustering to nodes. Vertex similarity is defined in terms of the similarity of their neighbourhood [10].

Structural equivalence: two nodes are structurally equivalent if they are connecting to the same set of actors. Structural equivalence is too restricted for practical use.

Example: nodes 1 and 3 are structurally equivalent; so are nodes 5 and 7.

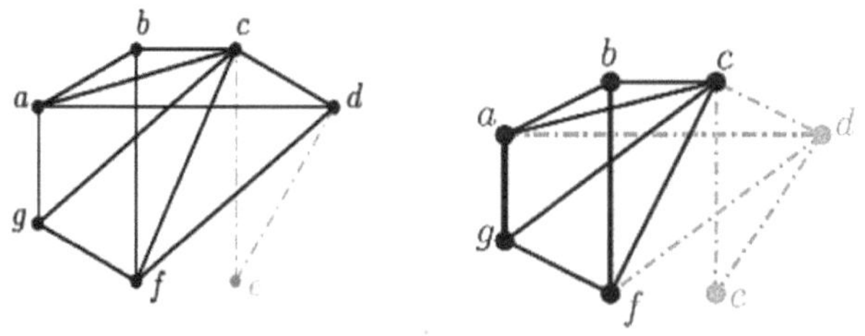

FIGURE 7.15 Maximal quasi-clique

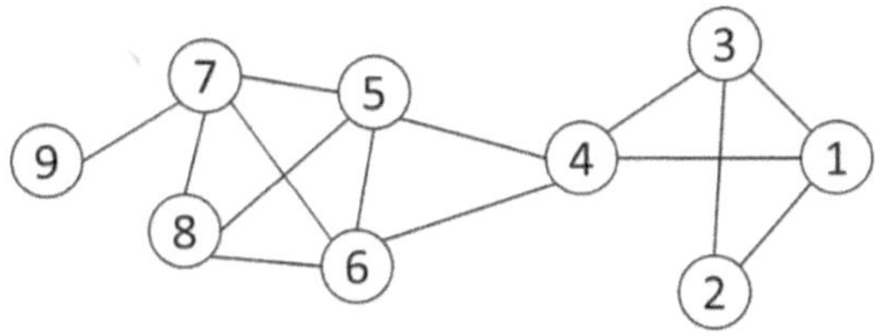

FIGURE 7.16 Example graph for structurally equivalent nodes or subgraphs

Jaccard similarity:

Jaccard $(vi, vj) = \frac{|Ni \cap Nj|}{|Ni \cup Nj|}$, where N_i and N_j are neighbour of node i and j.

$$J(5, 8) = \frac{|\{6\}|}{|\{1, 2, 6, 13\}|} = \frac{1}{4}$$

Cosine similarity:

$$\cos(\theta) = \frac{A.B}{\|A\| \, \|B\|} = \frac{\sum_{i=1}^{n} A_i X B_i}{\sqrt{\sum_{i=1}^{n} A_i^2} \sqrt{\sum_{i=1}^{n} B_i^2}}$$

where θ is the angle between two non-zero vectors A and B. We take the dot product of vectors A and B and divide it by the product of the magnitude of both vectors. Each i represents a dimension of the vectors; in this context, it could be letters or words. If there's no matching letter or word, then we disregard those (i.e., consider them as 0).

$$\text{cosine}(5, 8) = \frac{1}{\sqrt{2} \times \sqrt{3}} = \frac{1}{\sqrt{6}}$$

This similarity score ranges from 0 *to* 1, with 0 being the lowest (the least similar) and 1 being the highest (the most similar). For practical use with huge networks, consider the connections as features, use cosine or Jaccard similarity to compute vertex similarity and then apply the classical k-means clustering algorithm to detect

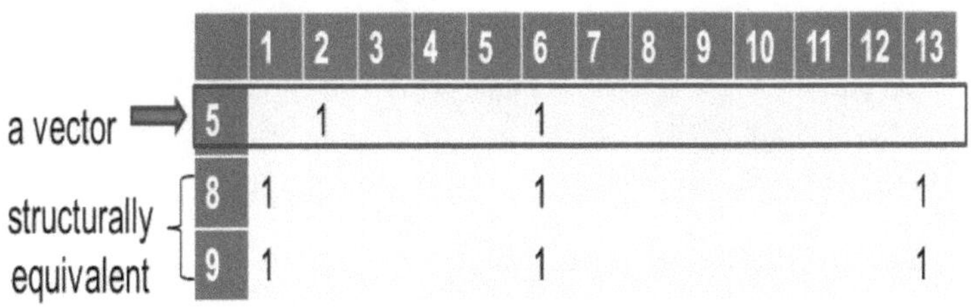

FIGURE 7.17 Node 8 and 9 are connected to same set of actors

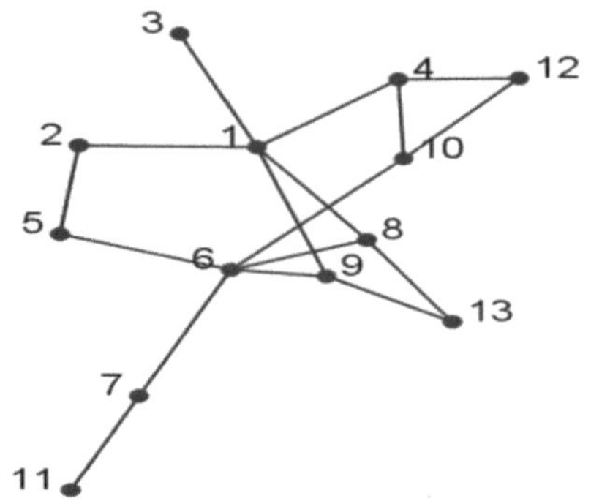

FIGURE 7.18 Graph for similarity example

communities. In the *k*-means clustering algorithm, each cluster is associated with a centroid (centre point) and each node is assigned to the cluster with the closest centroid [11].

BASIC *K*-MEANS ALGORITHM

1. Select *k* points as the initial centroids
2. Repeat from *k* clusters by assigning all points to the closest centroid.
3. Recompute the centroid of each cluster.
4. Until the centroid does not change.

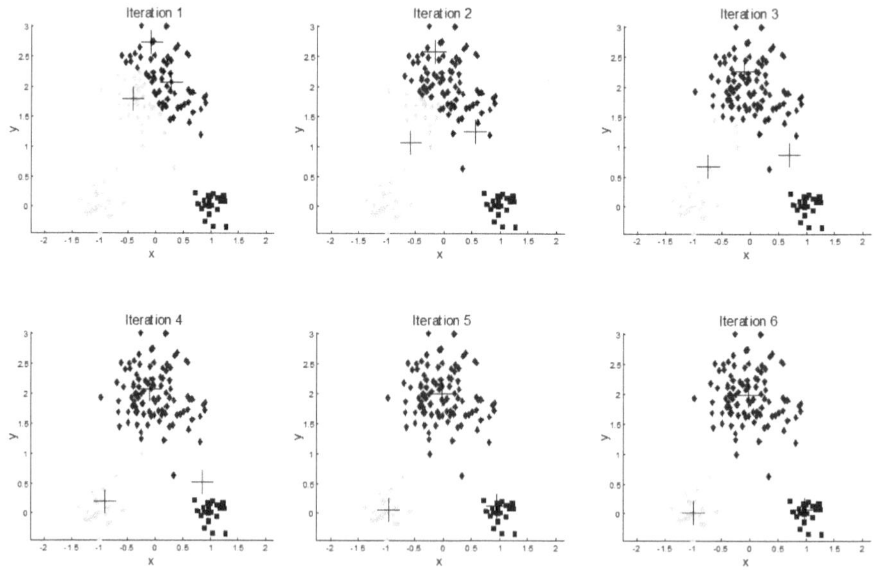

FIGURE 7.19 Illustration of *k*-means clustering

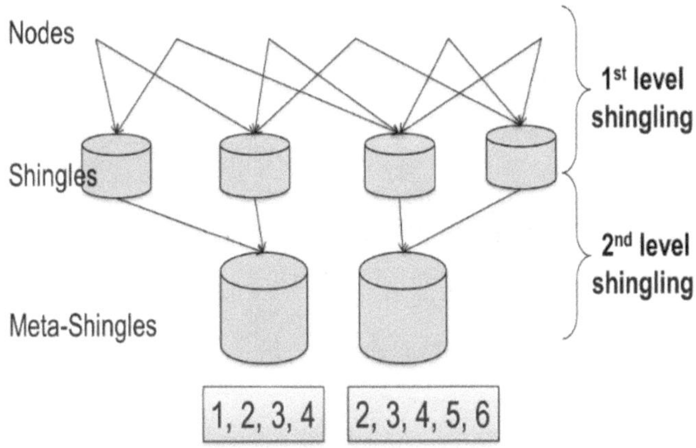

FIGURE 7.20 Two-level shingling

Shingling: the pair-wise computation of similarity can be time-consuming with millions of nodes. Thus, shingling can be exploited. In this, mapping each vector into multiple shingles is done so that the Jaccard similarity between two vectors can be computed by comparing the shingles. It is implemented using a quick hash function. Similar vectors share more shingles after transformation. Nodes of the same shingle can be considered belonging to one community. In reality, we can apply two-level shingling.

Latent-Space Models

Latent-space models transform the nodes in a network into a lower-dimensional space such that the distance or similarities between nodes are kept in the Euclidean space, then apply k-means clustering.

A classical multidimensional scaling (MDS) algorithm typically involves some linear algebra. Given a network, construct a proximity matrix to denote the distance between nodes (e.g., geodesic distance). The classical MDS algorithm rests on the fact that the coordinate matrix X can be derived by eigenvalue decomposition from the scalar product matrix $\boldsymbol{B} = \boldsymbol{XX'}$. The problem of constructing $\boldsymbol{B}$ from the proximity matrix $\boldsymbol{P}$ is solved by multiplying the squared proximities with the matrix $\boldsymbol{J} = \boldsymbol{I} - \boldsymbol{n}^{-1}\mathbf{11}'$. This procedure is called double centring [12].

The following steps summarize the algorithm of classical MDS:

1. Set up the matrix of squared proximities $\boldsymbol{P}^{(2)} = [\boldsymbol{p}^2]$.
2. Apply the double centring: $\boldsymbol{B} = \frac{1}{2}\boldsymbol{JP}^{(2)}\boldsymbol{J}$ using the matrix $\boldsymbol{J} = \boldsymbol{I} - \boldsymbol{n}^{-1}\mathbf{11}'$, where n is the number of objects.
3. Extract the m largest positive eigenvalues $\lambda_1, \ldots, \lambda_m$ of B and the corresponding m eigenvectors $\boldsymbol{e}_1, \ldots, \boldsymbol{e}_m$.
4. An m-dimensional spatial configuration of the n objects is derived from the coordinate matrix $\boldsymbol{X} = \boldsymbol{E}_m \boldsymbol{\Lambda}_m^{1/2}$, where $\boldsymbol{E}_m$ is the matrix of $\boldsymbol{m}$ eigenvectors and $\boldsymbol{\Lambda}_m$ is the diagonal matrix of $\boldsymbol{m}$ eigenvalues of B, respectively. Apply k-means to S to obtain clusters.

Example: Cities in Denmark

In order to illustrate classical MDS, assume that we have measured the distances between Copenhagen (cph), Arhus (aar), Odense (ode) and Aalborg (aal) on a map.

Therefore, the proximity matrix (showing the distances in millimetres) might look like:

	cph	aar	ode	aal
Cph	0	93	82	133
Aar	93	0	52	60
Ode	82	52	0	111
Aal	133	60	111	0

The matrix of squared proximities is:

$$P^{(2)} = \begin{bmatrix} 0 & 8649 & 6724 & 17689 \\ 8649 & 0 & 2704 & 3600 \\ 6724 & 2704 & 0 & 12321 \\ 17689 & 3600 & 12321 & 0 \end{bmatrix}$$

Since there are $n = 4$ objects, the matrix J is calculated by

$$J = \begin{bmatrix} 1 & 0 & 0 & 0 \\ 0 & 1 & 0 & 0 \\ 0 & 0 & 1 & 0 \\ 0 & 0 & 0 & 1 \end{bmatrix} - 0.25 \times \begin{bmatrix} 1 & 1 & 1 & 1 \\ 1 & 1 & 1 & 1 \\ 1 & 1 & 1 & 1 \\ 1 & 1 & 1 & 1 \end{bmatrix} = \begin{bmatrix} 0.75 & -0.25 & -0.25 & -0.25 \\ -0.25 & 0.75 & -0.25 & -0.25 \\ -0.25 & -0.25 & 0.75 & -0.25 \\ -0.25 & -0.25 & -0.25 & 0.75 \end{bmatrix}$$

Applying J to $P^{(2)}$ yields the double centred matrix B

$$\boldsymbol{B} = \tfrac{1}{2}\boldsymbol{J}\boldsymbol{P}^{(2)}\boldsymbol{J} = \begin{bmatrix} 5035.0625 & 1553.0625 & 258.9375 & 3740.9375 \\ 1553.0625 & 507.8125 & 5.3125 & 1039.9375 \\ 258.9375 & 5.3125 & 2206.8125 & 2471.0625 \\ 3740.9375 & 1039.9375 & 2471.0625 & 5172.062 \end{bmatrix}$$

For a two-dimensional representation of the four cities, the first two largest eigenvalues and the corresponding eigenvectors of $\boldsymbol{B}$ have to be extracted

$$\lambda_1 = 9724.168; \lambda_2 = 3160.986,\ e1 = \begin{pmatrix} 0.637 \\ 0.187 \\ 0.253 \\ 0.704 \end{pmatrix},\ e2 = \begin{pmatrix} 0.586 \\ 0.214 \\ 0.706 \\ 0.334 \end{pmatrix}$$

Finally, the coordinates of the cities (up to rotations and reflections) are obtained by multiplying eigenvalues and eigenvectors

$$X = \begin{bmatrix} -0.637 & 0.586 \\ 0.187 & 0.214 \\ -0.253 & 0.706 \\ 0.704 & 0.334 \end{bmatrix} \begin{bmatrix} \sqrt{9724.168} & 0 \\ 0 & \sqrt{3160.986} \end{bmatrix} = \begin{bmatrix} -62.831 & -32.97448 \\ 18.403 & 12.02697 \\ 24.960 & 39.71091 \\ 69.388 & 18.76340 \end{bmatrix}$$

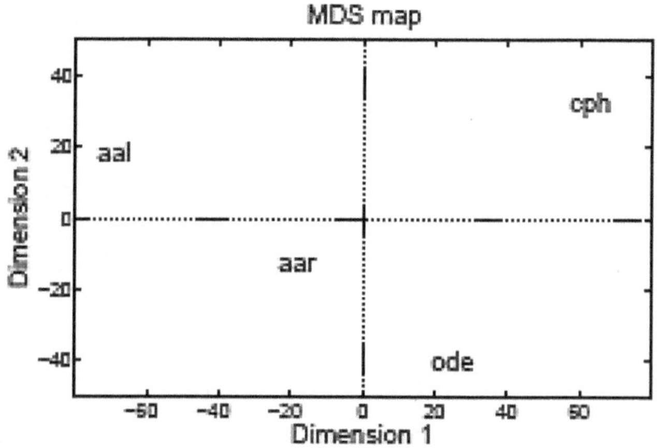

FIGURE 7.21 MDS map of the example

A classical MDS representation of the four Danish cities on a two-dimensional plane is shown in Figure 7.21.

The objective of block models is to minimize the difference between an interaction matrix and a block structure.

$$\min_{S,\Sigma} \left\| A - S\Sigma\, S^T \right\|_F$$

s. t. $S \in \{0,1\}^{n\times k}$, $\Sigma \in R^{k\times k}$ is diagonal.
Challenge: S is discrete, difficult to solve
Relaxation: allow S to be continuous satisfying $S^T S = I_k$
Solution: the top eigenvectors of A.
Post-processing: apply k-means to S to find the partition.

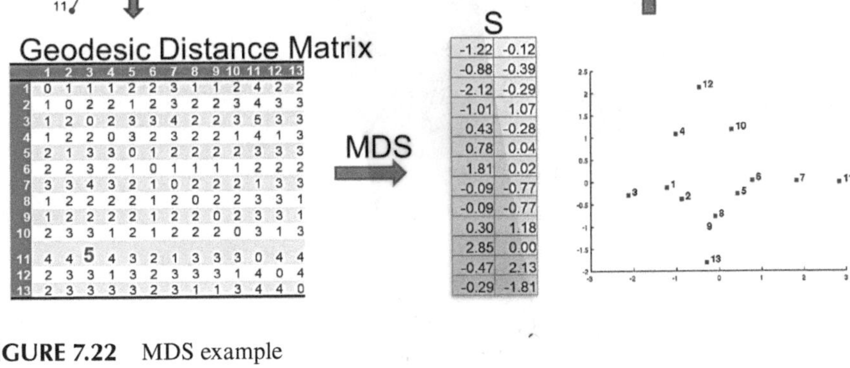

FIGURE 7.22 MDS example

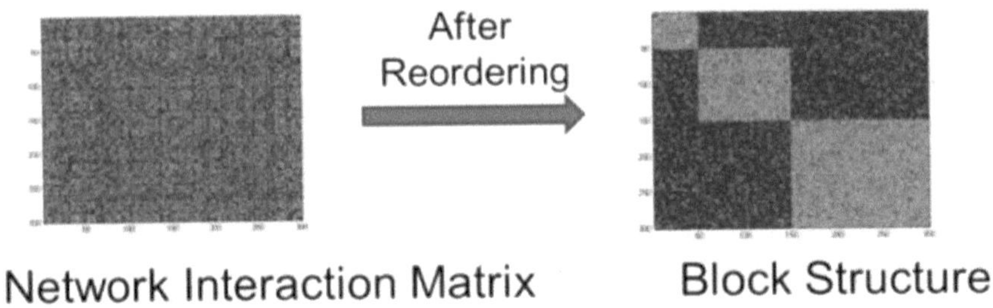

FIGURE 7.23 Block models

Cut-minimization: it can be observed that most interactions are within group whereas interactions between groups are few. Thus, the communities can be detected using the minimum cut problem. Cut is a partition of vertices of a graph into two disjoint sets and to find a graph partition such that the number of edges between the two sets is minimized is known as minimum cut problem.

$$\operatorname{cut}(C_1, C_2, \ldots, C_k) = \sum_{i=1}^{k} \operatorname{cut}(C_i, \bar{C}_i)$$

The limitation of the minimum cut problem is that it often finds communities of only one node (often returns an imbalanced partition, with one set being a singleton) thus, we need to change the objective function to consider community size k.

Ratio cut and normalized cut:

$$\text{Ratio cut } (\pi) = 1/\text{k} \sum_{i=1}^{k} \frac{\operatorname{cut}(C_i, \bar{C}_i)}{|C_i|}$$

$$\text{Normalized cut } (\pi) = 1/\text{k} \sum_{i=1}^{k} \frac{\operatorname{cut}(C_i, \bar{C}_i)}{\operatorname{vol}|C_i|}$$

where,
C_i: a community
$|C_i|$: number of nodes in C_i
$\operatorname{vol}|C_i|$: sum of degrees in C_i

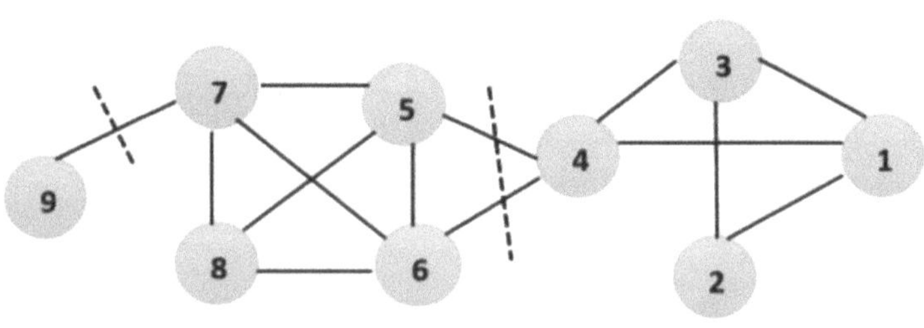

FIGURE 7.24 Example of a ratio cut

For partition in dots: π_1

$$\text{Ratio cut}\left(\pi_1\right)=\frac{1}{2}\left(\frac{1}{1}+\frac{1}{8}\right)=\frac{9}{16}=0.56$$

$$\text{Normalized cut}\left(\pi_1\right)=\frac{1}{2}\left(\frac{1}{1}+\frac{1}{27}\right)=\frac{14}{27}=0.52$$

For partition in dashes: π_2

$$\text{Ratio cut}\left(\pi_2\right)=\frac{1}{2}\left(\frac{2}{4}+\frac{2}{5}\right)=\frac{9}{20}=0.45<\text{Ratio cut}\left(\pi 1\right)$$

$$\text{Normalized cut}\left(\pi_2\right)=\frac{1}{2}\left(\frac{2}{12}+\frac{2}{16}\right)=\frac{7}{48}=0.15<\text{Normalized cut}\left(\pi 1\right)$$

Spectral Clustering

Both ratio cut and normalized cut can be reformulated as

$$\min_{S\in R^{n\times k}} T_r\left(S^T LS\right) s.\ t.\ S^T S=I$$

where,

$$L=D-A$$

$$\text{normalized}-L=I-D^{-\frac{1}{2}}AD^{-\frac{1}{2}}$$

$$D=\begin{matrix} d_1 & 0 & . & . & . & 0 \\ 0 & d_2 & . & . & . & 0 \\ . & . & . & . & . & . \\ . & . & . & . & . & . \\ . & . & . & . & . & . \\ 0 & 0 & . & . & . & d_n \end{matrix}$$

where, $D=\text{diag}(d_1, d_2,\ldots, d_n)$ and A is the diagonal matrix of degrees.
Solution: S are the eigenvectors of L with smallest eigenvalues (except the first one).
Post-processing: apply k-means to S also known as spectral clustering.

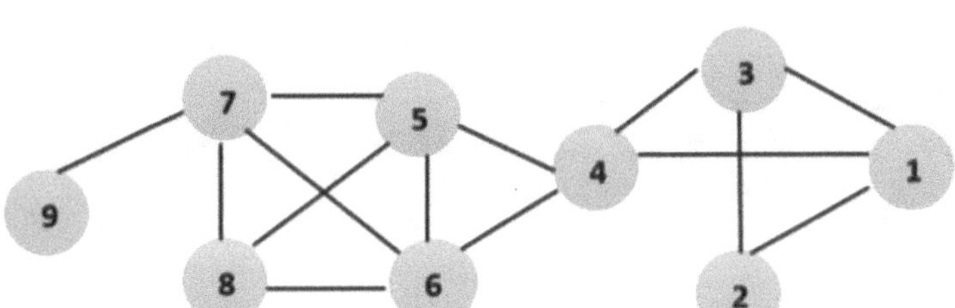

FIGURE 7.25 Example of spectral clustering

Spectral clustering example:

$$D = \text{diag}(3, 2, 3, 4, 4, 4, 4, 3, 1)$$

$$L = D - A = \begin{matrix} 3 & -1 & -1 & -1 & 0 & 0 & 0 & 0 & 0 \\ -1 & 2 & -1 & 0 & 0 & 0 & 0 & 0 & 0 \\ -1 & -1 & 3 & -1 & 0 & 0 & 0 & 0 & 0 \\ -1 & 0 & -1 & 4 & -1 & -1 & 0 & 0 & 0 \\ 0 & 0 & 0 & -1 & 4 & -1 & -1 & -1 & 0 \\ 0 & 0 & 0 & -1 & -1 & 4 & -1 & -1 & 0 \\ 0 & 0 & 0 & 0 & -1 & -1 & 4 & -1 & -1 \\ 0 & 0 & 0 & 0 & -1 & -1 & -1 & 3 & 0 \\ 0 & 0 & 0 & 0 & 0 & 0 & -1 & 0 & 1 \end{matrix}$$

The first eigenvector means all nodes belong to the same cluster, no use.

$$S = \begin{matrix} 0.33 & 0.38 \\ 0.33 & 0.48 \\ 0.33 & 0.38 \\ 0.33 & 0.12 \\ 0.33 & 0.16 \\ 0.33 & 0.16 \\ 0.33 & 0.30 \\ 0.33 & 0.24 \\ 0.33 & 0.51 \end{matrix}$$

S are the eigenvectors of L with smallest eigenvalues (except the first one). After applying k-means to S, we get two communities: $\{1, 2, 3, 4\}$ and $\{5, 6, 7, 8, 9\}$.

Graph modularity: modularity is a metric that measures the strength of division of a network into communities. Networks with high modularity have dense connections between the nodes within communities, but sparse connections between nodes in different communities. Modularity is often used in optimization methods for the detection of community structures.

Intuitively, it can be measured as the total number of community edges minus the expected number of edges in the absence of a community structure.

Relational network given by $G = (V, A)$

V : set of n vertices A : $n \times n$ adjacency matrix, m total edges.

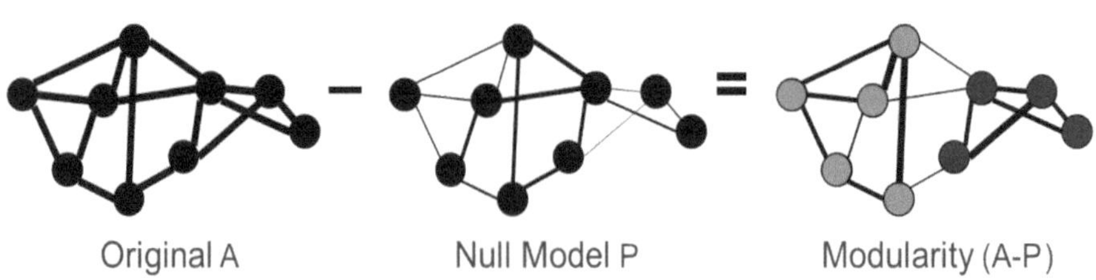

FIGURE 7.26 Graph modularity

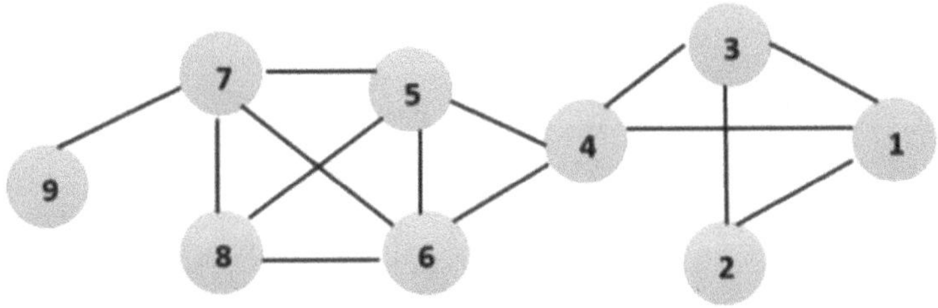

FIGURE 7.27 Example of modularity maximization

Modularity Maximization

Modularity measures the strength of a community partition by taking into account the degree distribution. Modularity measures the group interactions compared with the expected random connections in the group. In a network with m edges, for two nodes with degree d_i and d_j, expected random connections between them are $\frac{d_i\, d_j}{2m}$

For Figure 7.27, the expected number of edges between nodes 1 and 2 are:

$$3*2/(2*14)=3/14$$

Strength of a community:

$$\sum_{i\epsilon C, j\epsilon C} A_{ij} - \frac{d_i d_j}{2m}$$

Modularity matrix: $B = A - dd^T/2m$, that is, $B = A_{ij} - d_i d_j / 2m$

The interaction utility in a group: to partition the group into multiple groups, we maximize the modularity by

$$\text{Modularity } (Q) = \frac{1}{2m} \sum_{C} \sum_{i\epsilon C, j\epsilon C} A_{ij} - \frac{d_i d_j}{2m}$$

A larger value indicates a good community structure.

Optimal solution: top eigenvectors of the modularity matrix. Apply k-means to S as a post-processing step to obtain community partition.

Properties of modularity:

- Between $(-1, 1)$
- Modularity $= 0$ If all nodes are clustered into one group.
- Can automatically determine optimal number of clusters.
- Resolution limit of modularity.
- Modularity maximization might return a community consisting multiple small modules.

Limitation: Require the user to specify the number of communities beforehand.

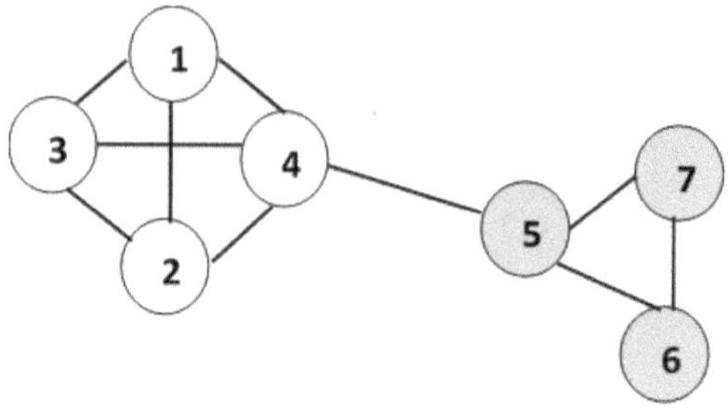

FIGURE 7.28 Example 1 of modularity maximization

Modularity maximization example 1:
Degrees of the seven nodes are of the graph in Figure 7.28:

$$d = \begin{bmatrix} 3 & 3 & 3 & 4 & 3 & 2 & 2 \end{bmatrix}$$

Total degree: $2m = 20$

i \ j	1	2	3	4	5	6	7
1	-0.45	0.55	0.55	0.4	-0.45	-0.3	-0.3
2	0.55	-0.45	0.55	0.4	-0.45	-0.3	-0.3
3	0.55	0.55	-0.45	0.4	-0.45	-0.3	-0.3
4	0.4	0.4	0.4	-0.8	0.4	-0.4	-0.4
5	-0.45	-0.45	-0.45	0.4	-0.45	0.7	0.7
6	-0.3	-0.3	-0.3	-0.4	0.7	-0.2	0.8
7	-0.3	-0.3	-0.3	-0.4	0.7	0.8	-0.2

Nodes 1, 2, 3 and 4 tend to form one community and nodes 5, 6 and 7 another. The modularity Q based on this division is the sum of all grey cells in the modularity matrix divided by $2m$: 0.355

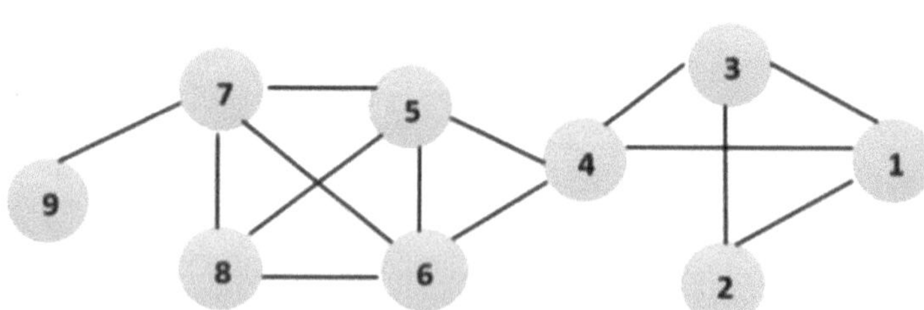

FIGURE 7.29 Example 2 of modularity maximization

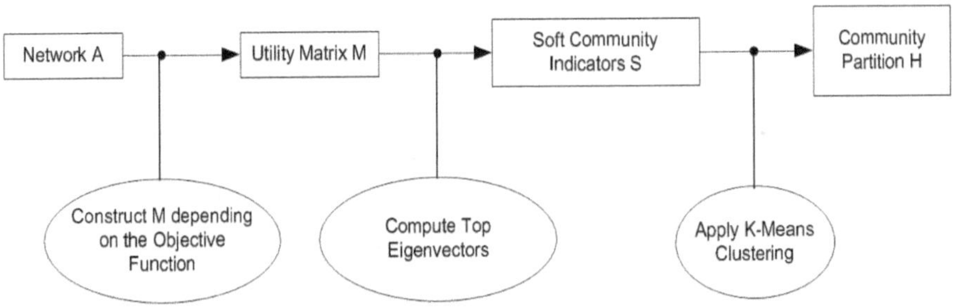

FIGURE 7.30 Unified view for community partition

Modularity maximization example 2:
Modularity matrix $B = A_{ij} - d_i d_j / 2m$

−0.32	0.79	0.68	0.57	−0.43	−0.43	−0.43	−0.32	−0.11
0.79	−0.14	0.79	−0.29	−0.29	−0.29	0.29	−0.21	−0.07
0.68	0.79	−0.32	0.57	−0.43	−0.43	−0.43	−0.32	−0.11
0.57	−0.29	0.57	−0.57	0.43	0.43	−0.57	−0.43	−0.14
−0.43	−0.29	−0.43	0.43	−0.57	0.43	0.43	0.57	−0.14
−0.43	−0.29	−0.43	0.43	0.43	−0.57	0.43	0.57	−0.14
−0.43	−0.29	−0.43	−0.57	0.43	0.43	−0.57	0.57	0.86
−0.32	−0.21	−0.32	−0.43	0.57	0.57	0.57	−0.32	−0.11
−0.11	−0.07	−0.11	−0.14	−0.14	−0.14	0.86	−0.11	−0.04

$$S = \begin{matrix} 0.44 & 0.00 \\ 0.38 & 0.23 \\ 0.44 & -0.00 \\ 0.17 & -0.48 \\ -0.29 & -0.32 \\ -0.29 & -0.32 \\ -0.38 & 0.34 \\ -0.34 & -0.08 \\ -0.14 & 0.63 \end{matrix}$$

After applying k-means two communities: $\{1, 2, 3, 4\}$ *and* $\{5, 6, 7, 8, 9\}$

7.2.4 Hierarchy-Centric Community Detection

This is to construct a hierarchical structure of communities. The goal is to build a hierarchical structure of communities based on network topology. This allows the analysis of a network at different resolutions. Hierarchical clustering is the most popular and widely used method to analyze social network data. In this method, nodes are compared with one another based on their similarity. Larger groups are built by joining groups of nodes based on their similarity. A criterion is introduced

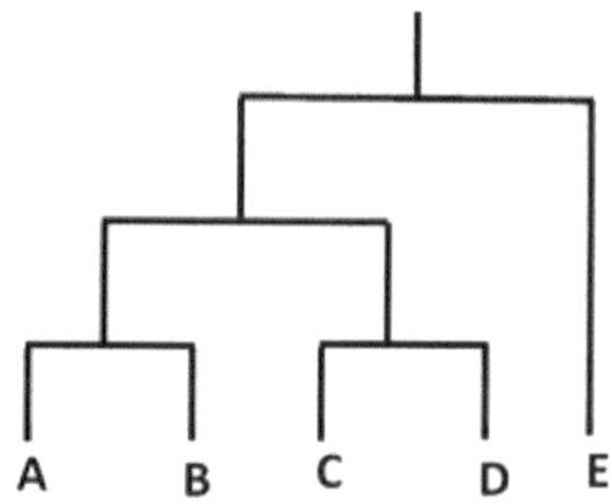

FIGURE 7.31 Dendrogram

to compare nodes based on their relationship. There are two types of hierarchical clustering approaches: agglomerative approach and divisive approach

Agglomerative Approach

This method is also called a bottom-up approach. In this method, each node represents a single cluster at the beginning; eventually, nodes start merging based on their similarities and all nodes belong to the same cluster. The user can specify termination condition, as the desired number of clusters. Output is a dendrogram, which can be represented as a set of order triples $\langle d, k, K \rangle$ where d is the threshold distance, k is the number of clusters and K is the set of clusters [13, 14, 15, 16, 17, 18].

3	<3, 1, {{A, B, C, D}}>
2	<2, 2, {{A, B, C, D}{E}}>
1	<1, 3, {{A, B}{C, D}{E}}>
0	<0, 5, {{A}{B}{C}{D}{E}}>

Agglomerative algorithm steps:

1. Start by assigning each item to a cluster, For N items, there will N clusters, each containing just one item.
2. Find the closest (most similar) pair of clusters and merge them into a single cluster, so that now we have one cluster less.
3. Compute distances (similarities) between the new cluster and each of the old clusters.
4. Repeat steps 2 and 3 until all items are clustered into a single cluster of size N.

Step 3 can be done in different ways: single-linkage, complete-linkage and average-linkage clustering [19, 20].

Single-linkage clustering (connectedness or minimum method): we consider the distance between one cluster and another cluster to be equal to the shortest distance from any member of one cluster to any member of the other cluster.

Single link: $L(R,S) = \min\big(D(i,j)\big),\ i \in R,\ j \in S$

Advantages: loose clusters, can handle non-elliptical shape

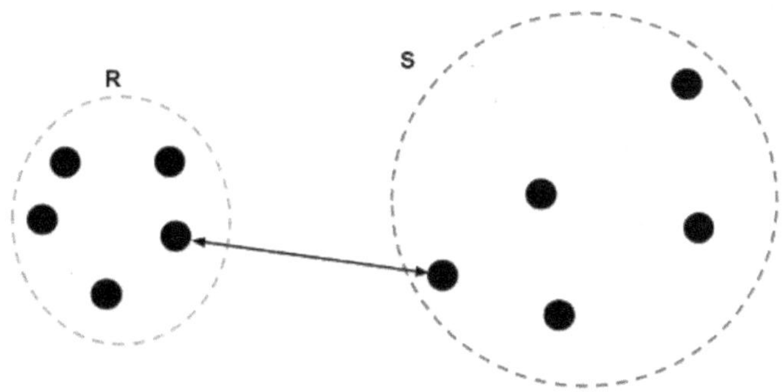

FIGURE 7.32 Single-linkage clustering

Disadvantages: sensitive to noise and outliers

Complete-linkage clustering (diameter or maximum method): we consider the distance between one cluster and another cluster to be equal to the greatest distance from any member of one cluster to any member of the other cluster.

Complete link: $L(R,S)=\max\left(D(i,j)\right),\ i\in R,\ j\in S$

Advantages: tight clusters, less susceptible to noise and outliers.

Disadvantages: biased towards global clusters.

Average-linkage clustering: we consider the distance between one cluster and another cluster to be equal to the average distance from any member of one cluster to any member of the other cluster.

Average link: $L(R,S)=\frac{1}{n_R+n_S}\sum_{i=1}^{n_R}\sum_{j=1}^{n_S}\min\left(D(i,j)\right),\ i\in R,\ j\in S$

where,

n_R – Number of data-points in R

n_S – Number of data-points in S

Advantages: in-between loose and tight clusters, less susceptible to noise and outliers.

Disadvantages: biased towards global clusters.

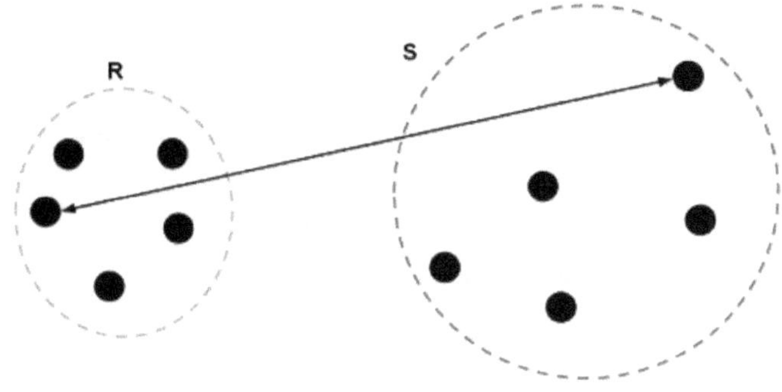

FIGURE 7.33 Complete-linkage clustering

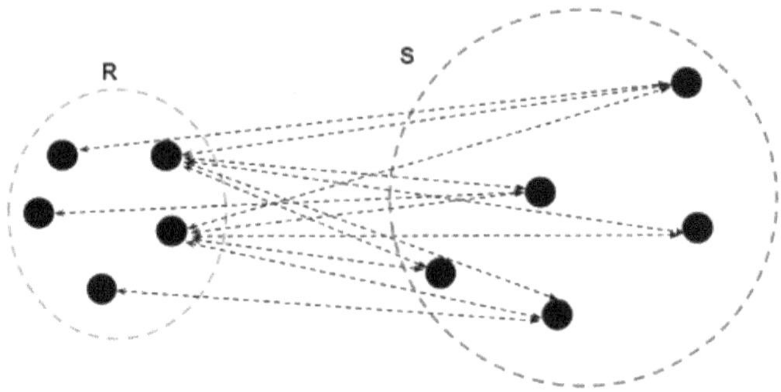

FIGURE 7.34 Average-linkage clustering

Agglomerative algorithm example:

Apply agglomerative with single, complete and average linkage on the following data:

1	0				
2	2	0			
3	6	3	0		
4	10	9	7	0	
5	9	8	5	4	0
	1	**2**	**3**	**4**	**5**

Note: for any type of linkage, always pick up the minimum distance from the matrix. We require to keep two adjacency matrices. Do not change to original matrix.

1	0				
2	2	0			
3	6	3	0		
4	10	9	7	0	
5	9	8	5	4	0
	1	**2**	**3**	**4**	**5**

Single linkage example: **minimum function**

In the original matrix, 1 and 2 are located close to each other at distance 2. Merge them into a single cluster.

1	0				
2	2	0			
3	6	3	0		
4	10	9	7	0	
5	9	8	5	4	0
	1	**2**	**3**	**4**	**5**

12	0			
3		0		
4		7	0	
5		5	4	0
	12	**3**	**4**	**5**

Min[(3,1),(3,2)]	Min[(4,1),(4,2)]	Min[(5,1),(5,2)]
Min[6,3]	Min[10,9]	Min[9, 8]
3	**9**	**8**

We will fill the blank grey cells with minimum values calculated in the above table.

12	0			
3	3	0		
4	9	7	0	
5	8	5	4	0
	12	**3**	**4**	**5**

Now, 12 and 3 are located close to each other at minimum distance 3. Merge them into a single cluster.

12	0			
3	3	0		
4	9	7	0	
5	8	5	4	0
	12	**3**	**4**	**5**

123	0		
4		0	
5		4	0
	123	**4**	**5**

Min[(4,12)(4,3)]	Min[(5,12)(5,3)]
Min[9,7]	Min[8,5]
7	**5**

We will fill the blank grey cells with minimum values calculated in the above table.

123	0		
4	7	0	
5	5	4	0
	123	**4**	**5**

123	0		
4	7	0	
5	5	4	0
	123	**4**	**5**

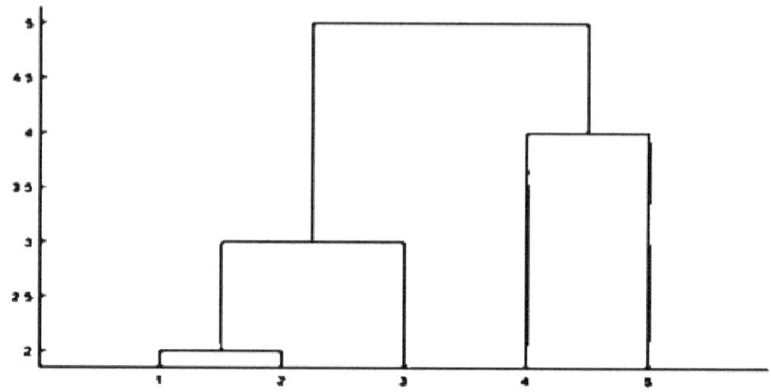

FIGURE 7.35 Dendrogram of a single-linkage example

Now, 4 and 5 are located close to each other at minimum distance 4. Merge them into a single cluster.

123	0	
45		0
	123	45

Min[(123,4)(123,5)]
Min[7,5]
5

We will fill the blank grey cells with minimum values calculated in the above table.

123	0	
45	5	0
	123	**45**

Thus, 123 and 45 are the clusters using single linkage.

Complete linkage example: maximum function

In the original matrix, 1 and 2 are located close to each other at distance 2. Merge them into a single cluster.

1	0				
2	2	0			
3	6	3	0		
4	10	9	7	0	
5	9	8	5	4	0
	1	**2**	**3**	**4**	**5**

12	0			
3		0		
4		7	0	
5		5	4	0
	12	**3**	**4**	**5**

max[(3,1),(3,2)]	max[(4,1),(4,2)]	max[(5,1),(5,2)]
max[6,3]	max[10,9]	max[9, 8]
6	**10**	**9**

We will fill the blank grey cells with maximum values calculated in the above table.

12	0			
3	6	0		
4	10	7	0	
5	9	5	4	0
	12	**3**	**4**	**5**

Now, 4 and 5 are located close to each other at minimum distance 4. Merge them into a single cluster.

12	0			
3	6	0		
4	10	7	0	
5	9	5	4	0
	12	**3**	**4**	**5**

12	0		
3	6	0	
45			0
	12	**3**	**45**

max[(4,12)(5,12)]	max[(4,3)(5,3)]
max[10,9]	max[7,5]
10	**7**

We will fill the blank grey cells with maximum values calculated in the above table.

12	0		
3	6	0	
45	10	7	0
	12	**3**	**45**

Now, 12 and 3 are located close to each other at minimum distance 6. Merge them into a single cluster.

12	0		
3	6	0	
45	10	7	0
	12	**3**	**45**

123	0	
45		0
	123	45

max[(12,45)(3,45)]
max[10,7]
10

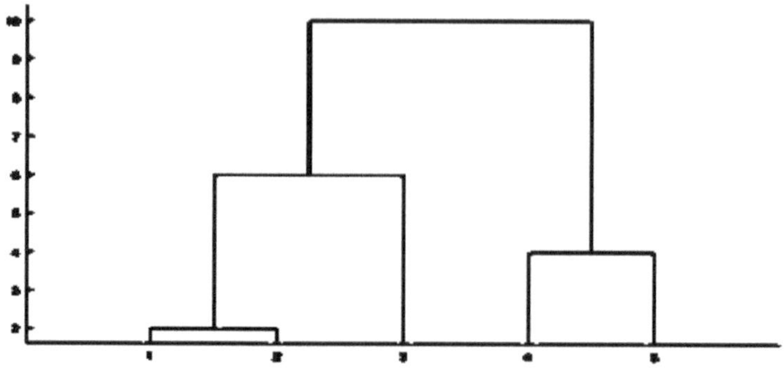

FIGURE 7.36 Dendrogram of a complete-linkage clustering example

We will fill the blank grey cells with maximum values calculated in the above table.

123	0	
45	10	0
	123	**45**

Thus, 123 and 45 are the clusters using complete linkage.

Average linkage example: average function

In the original matrix, 1 and 2 are located close to each other at distance 2. Merge them into a single cluster.

1	0				
2	2	0			
3	6	3	0		
4	10	9	7	0	
5	9	8	5	4	0
	1	**2**	**3**	**4**	**5**

12	0			
3		0		
4		7	0	
5		5	4	0
	12	**3**	**4**	**5**

(1,2)(3)	(1,2)(4)	(1,2)(5)
avg[(3,1),(3,2)]	avg[(4,1),(4,2)]	avg[(5,1),(5,2)]
avg[6,3]	avg[10,9]	avg[9, 8]
4.5	**9.5**	**8.5**

We will fill the blank grey cells with average values calculated in the above table.

12	0			
3	4.5	0		
4	9.5	7	0	
5	8.5	5	4	0
	12	**3**	**4**	**5**

Now, 4 and 5 are located close to each other at minimum distance 4. Merge them into a single cluster.

12	0			
3	4.5	0		
4	9.5	7	0	
5	8.5	5	4	0
	12	**3**	**4**	**5**

12	0		
3	4.5	0	
45			0
	12	**3**	**45**

(1,2) (4,5)	(3) (4,5)
avg[(4,1) (4,2)(5,1) (5,2)]	avg[(4,3)(5,3)]
avg[10,9,9,8]	avg[7,5]
9	**6**

We will fill the blank grey cells with average values calculated in the above table.

12	0		
3	4.5	0	
45	9	6	0
	12	**3**	**45**

Now, 12 and 3 are located close to each other at minimum distance 4.5. Merge them into a single cluster.

12	0		
3	4.5	0	
45	9	6	0
	12	**3**	**45**

123	0	
45		0
	123	**45**

(1,2,3)(4,5)
Avg(4,1)(4,2)(5,1)(5,2)(3,4)(3,5)]
max[10,9,9,8,7,5]
8

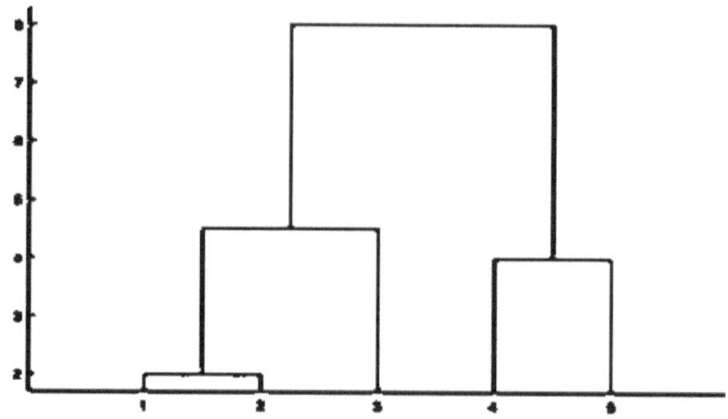

FIGURE 7.37 Dendrogram of an average-linkage clustering example

We will fill the blank grey cells with average values calculated in the above table.

123	0	
45	8	0
	123	**45**

Thus, 123 and 45 are the clusters using average linkage.

Divisive Approach

In divisive methods, we go the other way around. We start with the complete graph and take off the edges iteratively. The edge with the highest weight is removed first. At every step, the edge-weight calculation is repeated, since the weight of the remaining edges changes after an edge is removed. After a certain number of steps, we get clusters of densely connected nodes. This method is also called a top-down approach. Initially, all nodes belong to the same cluster; eventually, each node forms its own cluster [21]. The divisive approach is less widely used due to its complexity compared with the agglomerative approach.

Divisive algorithm: simple approach based on the minimum spanning tree (MST)

1. Compute an MST for the given adjacency matrix.
2. Repeat.
3. Create a new cluster by breaking the link corresponding to the largest distance until only a singleton cluster remains.

Example:

	A	B	C	D	E
A	0	1	2	2	3
B	1	0	2	4	3
C	2	2	0	1	5
D	2	4	1	0	3
E	3	3	5	3	0

From the adjacency matrix, create MST by Prim's or Kruskal's algorithm:

Edge	Cost
A-B	1
C-D	1
A-C	2
A-D	2
B-C	2
A-E	3
B-E	3
D-E	3
B-D	4
C-E	5

Edge	Cost	
A-B	1	
C-D	1	
A-C	2	
A-D	2	X
B-C	2	X
A-E	3	
B-E	3	X
D-E	3	X
B-D	4	X
C-E	5	X

The largest edge is between *A* & *E*. Cutting this edge results into two clusters $\{E\}$ and $\{A, B, C, D\}$. Next, remove the edge between *A* & *C*. This split creates three clusters $\{A, B\}$, $\{C, D\}$ and $\{E\}$. Next break *A* & *B*. This split creates four clusters $\{A\}$, $\{B\}$, $\{C, D\}$ and $\{E\}$. Next break *C* & *D*. This split creates five clusters $\{A\}$, $\{B\}$, $\{C\}$, $\{D\}$ and $\{E\}$.

Girvan and Newman's algorithm: we must find additional criteria to find which partition is the best one. Girvan and Newman offered an algorithm to solve the problems with spectral methods. It is based on the divisive method and hierarchical clustering. The divisive method repeatedly identifies and removes edges connecting

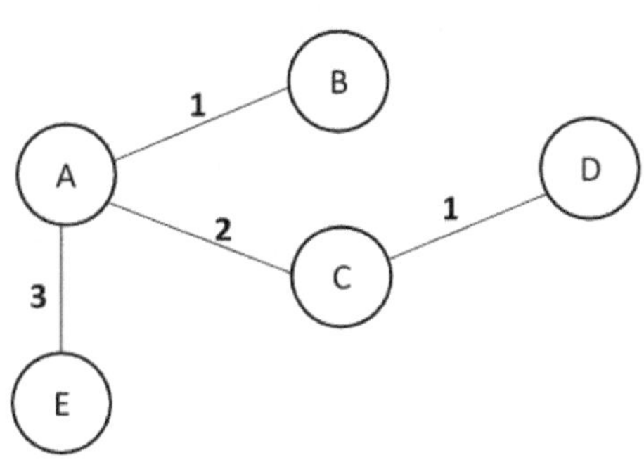

FIGURE 7.38 An MST of given example

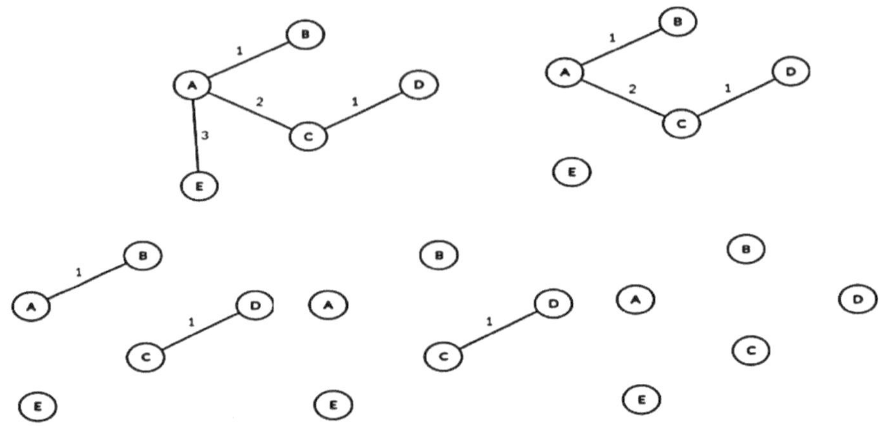

FIGURE 7.39 Forming clusters from an MST

densely connected regions. It uses edge betweenness that is the number of the shortest paths passing through the edge to identify edges to remove them. It also removes the links that connect clusters. The algorithm is designed to remove edges in decreasing betweenness [22].

Steps of Girvan and Newman's algorithm:

1. Compute betweenness for all links.
2. Remove the link with the highest betweenness.
3. Recompute betweenness for all remaining links.
4. Go to 2.

Girvan and Newman's algorithm example: for the given graph, calculate the betweenness centrality for the edges.

Let us start, we need to find shortest path first, so we are using BFS (breath first search).

First start with node A. In Figure 7.40, the adjacent nodes of node A are B and D and the adjacent nodes of B are C and E; likewise, D's adjacent E, E's adjacent is F and C's adjacent is F. Thus, this is the final graph that we get after taking node A from where we traverse all the nodes with the shortest path first.

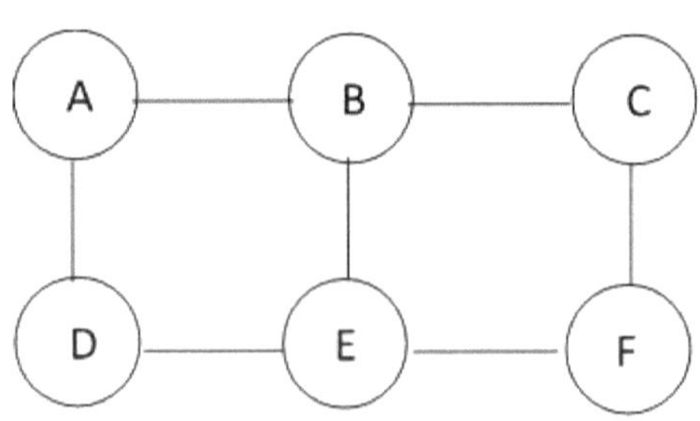

FIGURE 7.40 Girvan and Newman's algorithm example graph

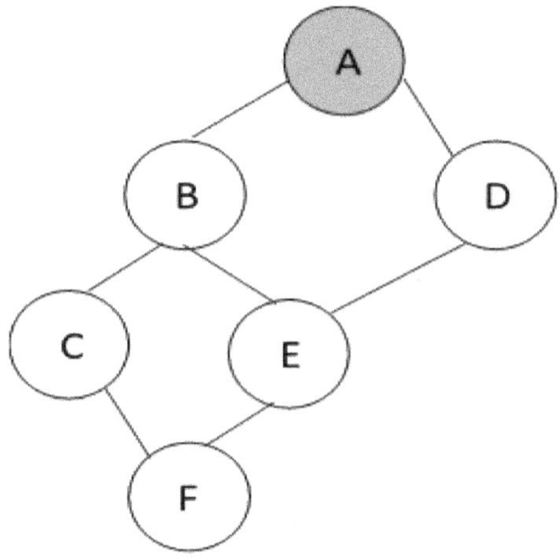

FIGURE 7.41 BFS traversal of the graph in Figure 7.40

Now let us calculate the values for the edges. A is the parent node; from A, we will give values to all the other nodes. A to B is direct path so the value of node B is 1; A to D is direct path so the value of node D is also 1. B to C can be travelled directly; the value of node C is 1; Node E can be travelled from B as well as node D and each is which is having value 1. Thus, E will have a weight of 1 + 1 = 2.

F can be traversed from nodes C and E; thus, F will have a weight 2 + 1 = 3; or we can say it will take 3 steps to reach F from A. Thus, Figure 7.42 is the graph where we have got the values of how much time it takes from A to various nodes.

Now we need to find the values for each edge. We will start from the bottom (i.e., from the node F). For F to C; F has value 3 and C has value 1. From F, we can reach either C or E; so it will be divided

For E to C =1/3 = 0.33 and for F to E=2/3 = 0.667

Now, C to B; C itself has value 1 thus the weight of edge CB will be 1 + 0.33 (lower edge value) = 1.33.

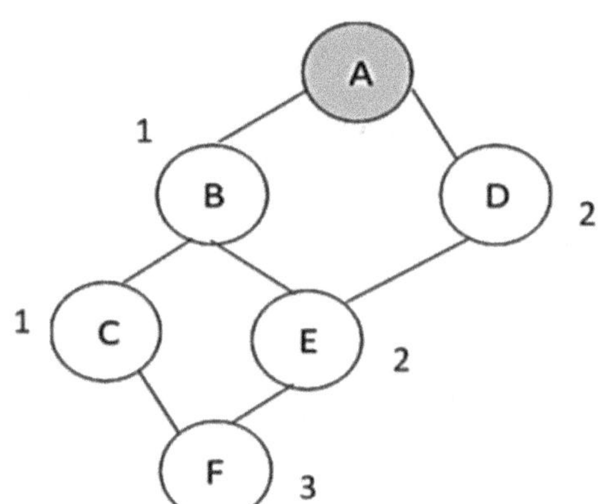

FIGURE 7.42 Weighted graph after BFS

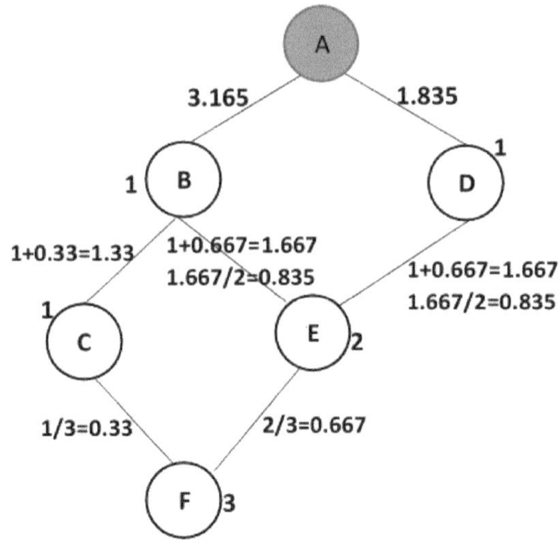

FIGURE 7.43 Edge weights after traversing from node A

For E to D = 1 + 0.667 (previous edge weight) = 1.667 and for E to B = 1 + 0.667 (previous edge weight) = 1.667. The value of E is getting diversify at two places thus the value we get here is being divided by 2. So, the final edge weight of EB and ED will be $E\ to\ B = \frac{1.667}{2} = 0.835$ and E to $D = \frac{1.667}{2} = 0.835$.

Now we will move further, the value of the edge B to $A = 1.33 + 1 + 0.835 = 3.165$ edge D to $A = 1 + 0.835 = 1.835$;

Note: By default, each node has weight of 1 and it will be added to the next edge weight.

So, Figure 7.43 is the complete graph for node A. We have traversed from A to all the different nodes and edges. The values calculated are placed.

Now, start from node B and repeat the above explained procedure.

Thus, this is the final graph that we get after taking node B from where we traverse all the nodes with the shortest path first.

So, the figure is the complete graph for node B. We have traversed from B to all the different nodes and edges. The values calculated are placed.

Now, start from node C and repeat the above explained procedure.

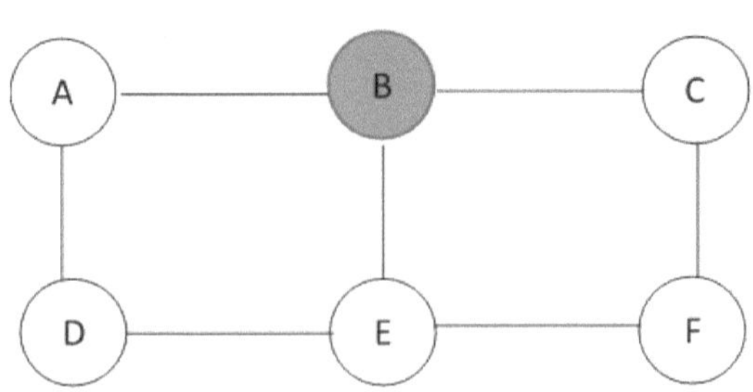

FIGURE 7.44 Input graph with source node B

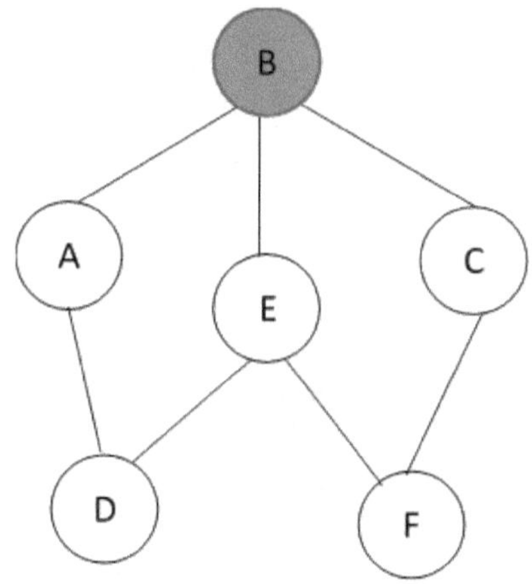

FIGURE 7.45 BFS traversal of the graph Figure 7.44

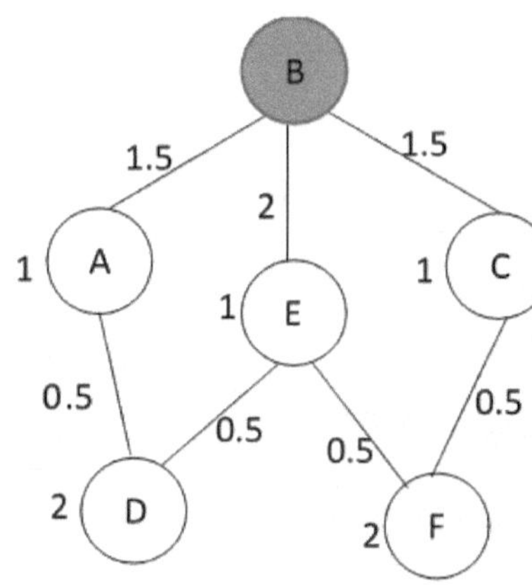

FIGURE 7.46 Edge weights after traversing from node B

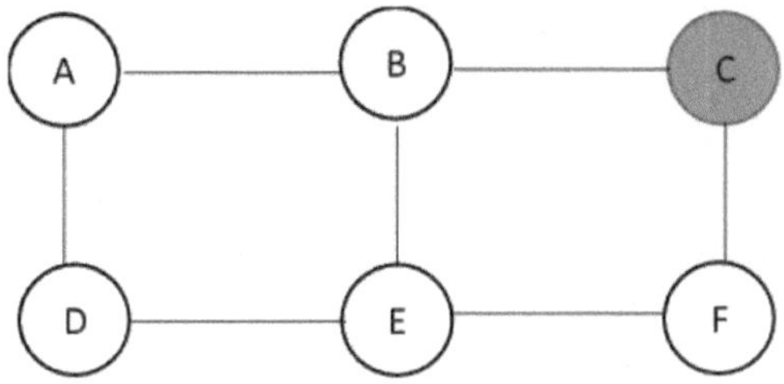

FIGURE 7.47 Input graph with source node C

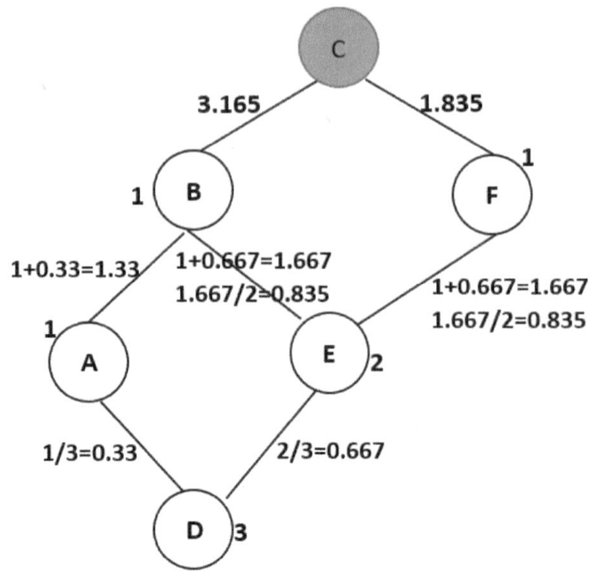

FIGURE 7.48 Edge weights after traversing from node C

So, Figure 7.48 is the complete graph for node C. We have traversed from C to all the different nodes and edges. The values calculated are placed.

Now, start from node D and repeat the above explained procedure.

So, Figure 7.50 is the complete graph for node D. We have traversed from D to all the different nodes and edges. The values calculated are placed.

Now, start from node E and repeat the above explained procedure.

So, Figure 7.52 is the complete graph for node E. We have traversed from E to all the different nodes and edges. The values calculated are placed.

Now, start from node F and repeat the above explained procedure.

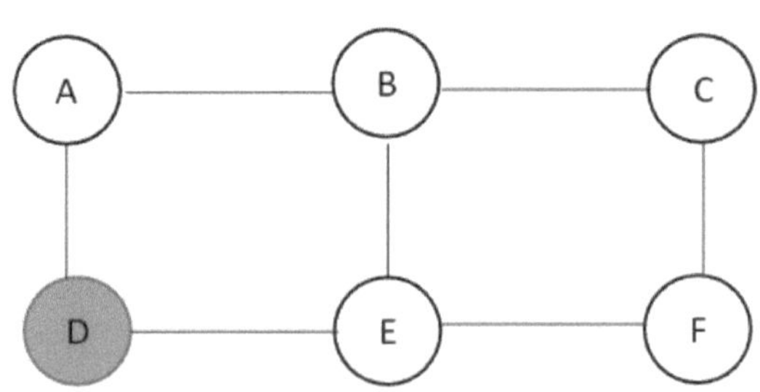

FIGURE 7.49 Input graph with source node D

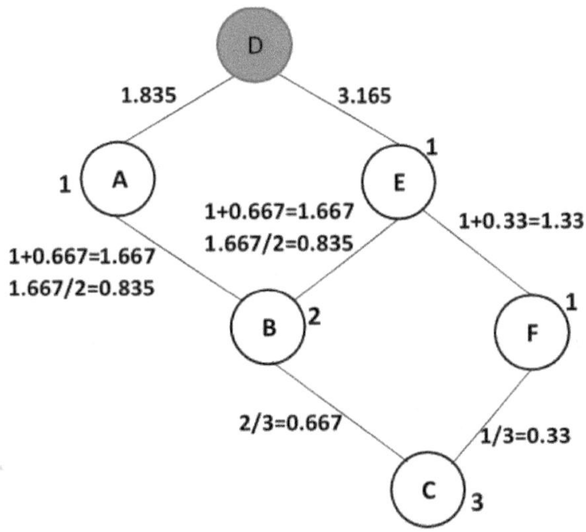

FIGURE 7.50 Edge weights after traversing from node D

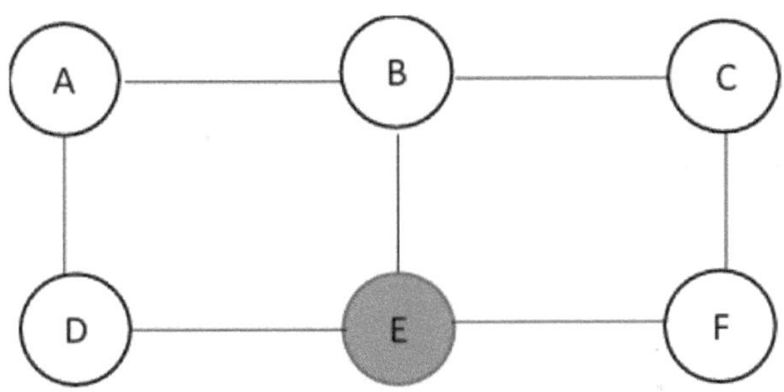

FIGURE 7.51 Input graph with source node E

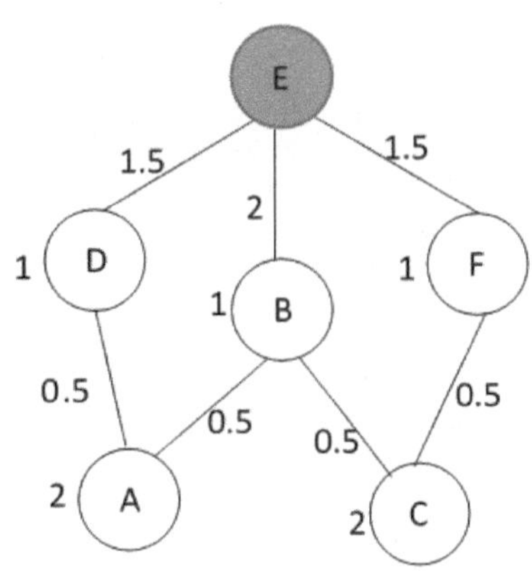

FIGURE 7.52 Edge weights after traversing from node E

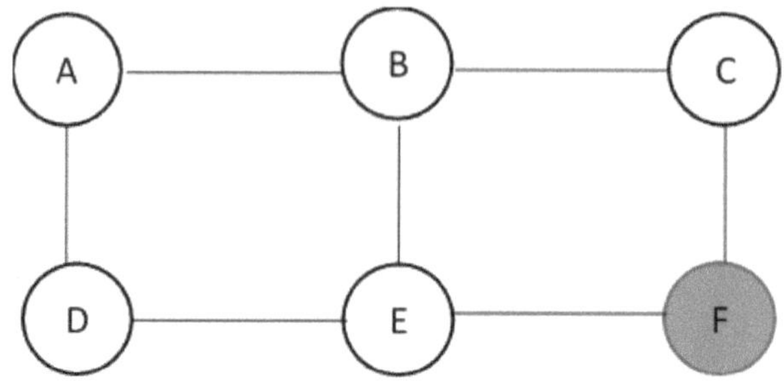

FIGURE 7.53 Input graph with source node F

So, Figure 7.54 is the complete graph for node F. We have traversed from F to all the different nodes and edges. The values calculated are placed.

Now, we will get the final weights of all the edges present in the graph by adding the shortest distance from each and every node that traverses the particular edge (Table 7.1).

Thus we will get the graph with final edge weights.

Now, remove the edges with minimum edge weights.

Thus {(A, B), (B, E), (C, F)} are the final communities after applying the edge betweenness procedure.

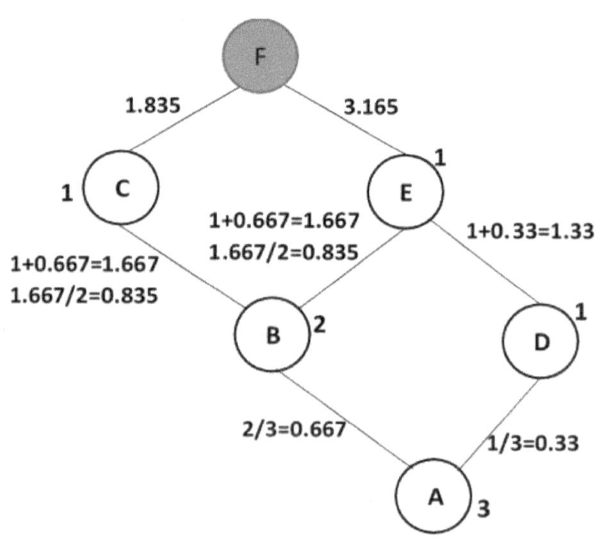

FIGURE 7.54 Edge weights after traversing from node F

TABLE 7.1
Edge Betweenness of All the Edges

Edges	Edge Betweenness
AB	3.165 + 1.5 + 1.33 + 0.835 + 0.5 + 0.667 = 8
AD	1.835 + 0.5 + 0.33 + 1.835 + 0.5 + 0.33 = 5.33
BC	3.165 + 1.5 + 1.33 + 0.835 + 0.5 + 0.667 = 8
BE	0.835 + 2 + 0.835 + 0.835 + 2 + 0.835 = 7.34
CF	1.835 + 0.5 + 0.33 + 1.835 + 0.5 + 0.33 = 5.33
DE	3.165 + 1.5 + 1.33 + 0.835 + 0.5 + 0.667 = 8
EF	3.165 + 1.5 + 1.33 + 0.835 + 0.5 + 0.667 = 8

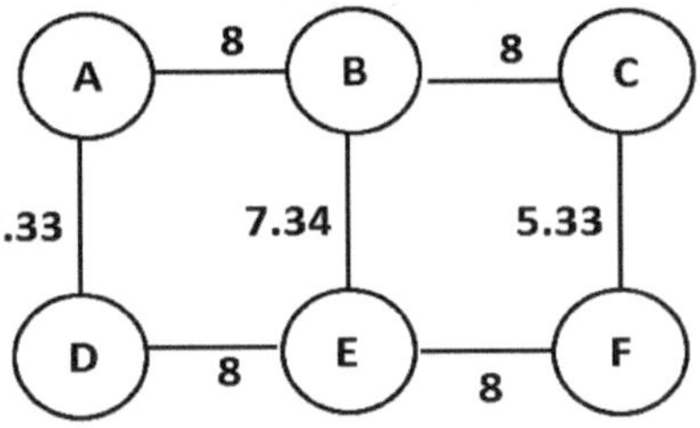

FIGURE 7.55 Graph with total edge betweenness

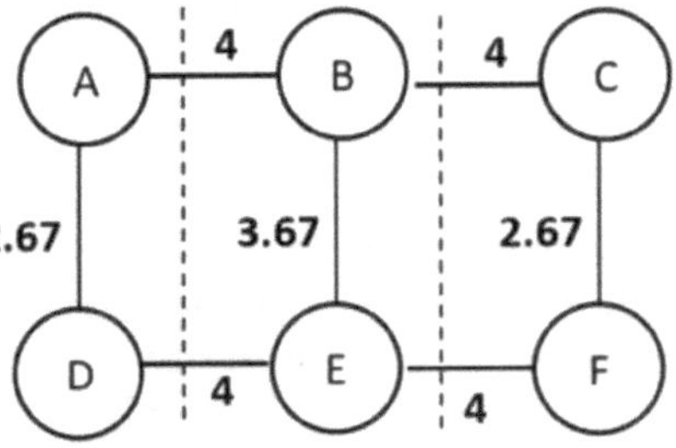

FIGURE 7.56 Selection of maximum weight edges

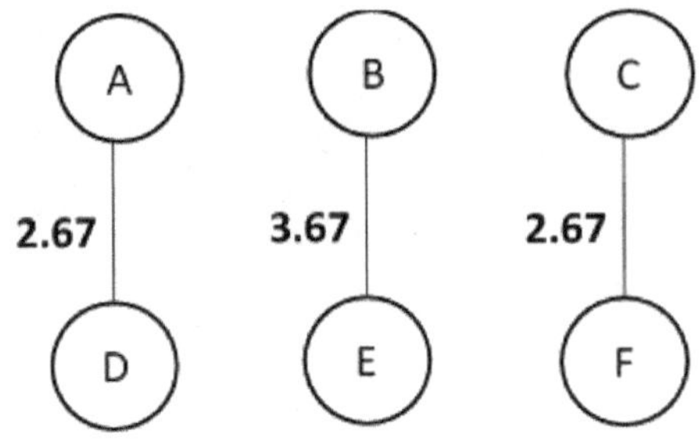

FIGURE 7.57 Removal of maximum weight edges

7.3 COMMUNITY EVALUATION

For groups with clear definitions – for example, cliques, k-cliques, k-clubs, quasi-cliques – evaluation of the communities verifies whether extracted communities satisfy the definition. For networks with ground truth information, get the normalized mutual information and accuracy of pairwise community memberships by comparing a clustering result with ground truth clustering result.

From the figure, it can be observed that the number of communities after grouping can be different from the ground truth. There is no clear community correspondence between clustering result and the ground truth. To resolve this, normalized mutual information (NMI) can be used.

Normalized Mutual Information

Entropy: the information contained in a distribution.

$$H(x) = \sum_{x \in X} p(x) \log p(x)$$

Mutual information: the shared information between two distributions.

$$I(X;Y) = \sum_{y \in Y} \sum_{x \in X} p(x,y) \log\left(\frac{p(x,y)}{p_1(x)\,p_2(y)}\right)$$

NMI: lies between 0 and 1.

$$NMI(X;Y) = \frac{I(X;Y)}{\sqrt{H(X)H(Y)}}$$

Consider a partition as a distribution (probability of one node falling into one community). We can compute the matching between two clusterings.

NMI: consider two partitions π^a and π^b with size $k^{(a)}$ and $k^{(b)}$.

$n_{h,l}$ is the number of vertices belonging to the hth community for (a) and to the lth community for (b). n_h^a is the number of vertices belong to the hth community for the partition (a) and n_l^b is the number of vertices belong to the lth community for the partition (b). Note that the entropy information gain is $0 <= NMI <= 1$.

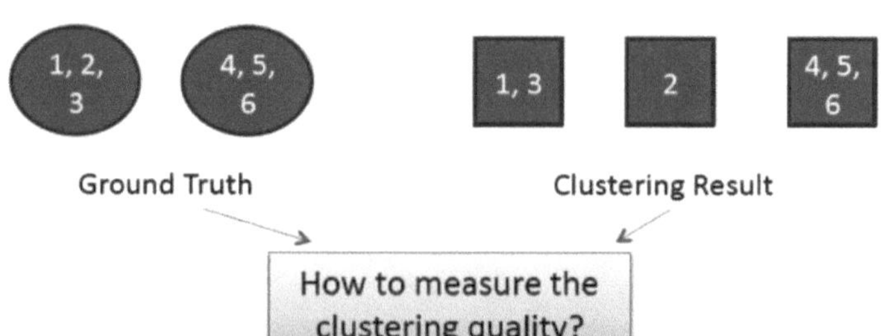

FIGURE 7.58 Ground truth clustering result and clustering result

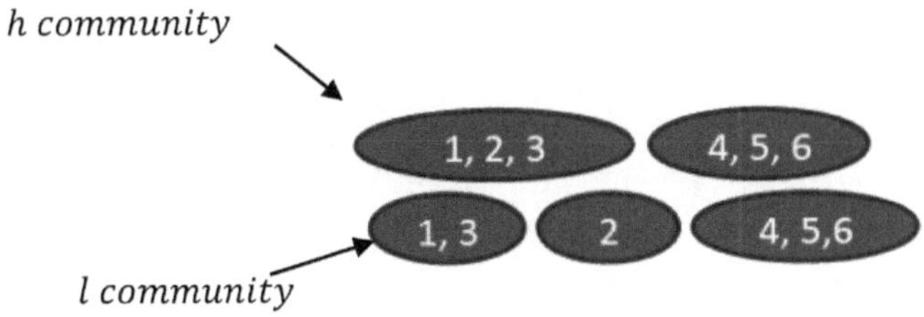

FIGURE 7.59 NMI example

Rewrite the equations:

$$H(x)=\sum_{x\in X}p(x)\log p(x)\Rightarrow H\left(\pi^a\right)=\sum_{h}^{k^{(a)}}\frac{n_h^a}{n}\log\left(\frac{n_h^a}{n}\right)\text{ and }H\left(\pi^b\right)=\sum_{l}^{k^{(b)}}\frac{n_l^b}{n}\log\left(\frac{n_l^b}{n}\right)$$

$$I(X;Y)=\sum_{y\in Y}\sum_{x\in X}p(x,y)\log\left(\frac{p(x,y)}{p_1(x)\,p_2(y)}\right)=I\left(\pi^a,\pi^b\right)=\sum_{h}\sum_{l}\frac{n_{h,l}}{n}\log\left(\frac{\frac{n_{h,l}}{n}}{\frac{n_h^a}{n}\cdot\frac{n_l^b}{n}}\right)$$

$$NMI(X;Y)=\frac{I(X;Y)}{\sqrt{H(X)H(Y)}}=NMI\left(\pi^a,\pi^b\right)$$

$$=\frac{\sum_{h=1}^{k^{(a)}}\sum_{l=1}^{k^{(b)}}n_{h,l}\log\left(\frac{n.n_{h,l}}{n_h^{(a)}.n_l^{(b)}}\right)}{\sqrt{\left(\sum_{h=1}^{k^{(a)}}n_h^{(a)}\log\left(\frac{n_h^a}{n}\right).\sum_{l=1}^{k^{(b)}}n_l^{(b)}\log\left(\frac{n_l^b}{n}\right)\right)}}$$

NMI example:

Partition a: $\pi^a=[1, 1, 1, 2, 2, 2]$
Partition b: $\pi^b=[1\text{–}3]$

$$n=6;\ k^{(a)}=2;\ k^{(b)}=3$$

	n_h^a
$h=1$	3
$h=2$	3

	n_l^b
$l=1$	2
$l=2$	1
$l=3$	3

$N_{h,l}$	$l=1$	$l=2$	$l=3$
$h=1$	2	1	0
$h=2$	0	0	3

$$NMI\left(\pi^a,\pi^b\right)=\frac{\sum_{h=1}^{k^{(a)}}\sum_{l=1}^{k^{(b)}} n_{h,l}\log\left(\frac{n.n_{h,l}}{n_h^{(a)}.n_l^{(b)}}\right)}{\sqrt{\left(\sum_{h=1}^{k^{(a)}} n_h^{(a)}\log\left(\frac{n_h^a}{n}\right).\sum_{l=1}^{k^{(b)}} n_l^{(b)}\log\left(\frac{n_l^b}{n}\right)\right)}}=0.8278$$

Accuracy of Pairwise Community Memberships

Consider all the possible pairs of nodes and check whether they reside in the same community. An error occurs if two nodes belonging to the **same** community are assigned to **different** communities after clustering and two nodes belonging to **different** communities are assigned to the **same** community.

Construct a contingency table:

		Ground Truth	
		$C(v_i)=C(v_j)$	$C(v_i)\neq C(v_j)$
Clustering Result	$C(v_i)=C(v_j)$	a	b
	$C(v_i)\neq C(v_j)$	c	d

$$\text{Accuracy}=\frac{a+d}{a+b+c+d}=\frac{a+d}{n(n-1)/2}$$

Accuracy example:

		Ground Truth	
		$C(v_i)=C(v_j)$	$C(v_i)\neq C(v_j)$
Clustering Result	$C(v_i)=C(v_j)$	4	0
	$C(v_i)\neq C(v_j)$	2	9

$$\text{Accuracy}=\frac{4+9}{4+2+9+0}=13/15$$

FIGURE 7.60 Accuracy example

FIGURE 7.61 An animal community and a health community

Community Evaluation using Semantics

This is done for the networks with semantics. Networks also come with semantic or attribute information of nodes or connections. Human subjects can verify whether the extracted communities are coherent. This kind of evaluation is qualitative. It is also intuitive and helps understand a community. Examples of semantic networks are shown in Figure 7.61.

Community Evaluation without Ground Truth

This is done for the networks without ground truth or semantic information. This is the most common situation. An option is to resort to cross validation, i.e., extract communities from a [training] network and evaluate the quality of the community structure on a network constructed from a different date or based on a related type of interaction [6, 7].

Quantitative evaluation functions are modularity and block model approximation error.

Modularity: once we have a network partition, we can compute its modularity. The method with high modularity wins.

Let e_{ij} be the fraction of edges in the network that connect nodes in community i to those in community j. Let $a_i = \sum_j e_{ij}$, then the modularity $Q = \sum_i \left(e_{ii} - a_i^2\right)$; values approaching $Q = 1$ indicate networks with strong community structure and values for real networks typically fall in the range from 0.3 *to* 0.7.

7.4 PROBLEMS FOR SELF-ASSESSMENT

Q1. For the graph below:

a. Find out the maximum clique (show all the steps).
b. Find overlapping communities using the CPM for cliques of size 4.

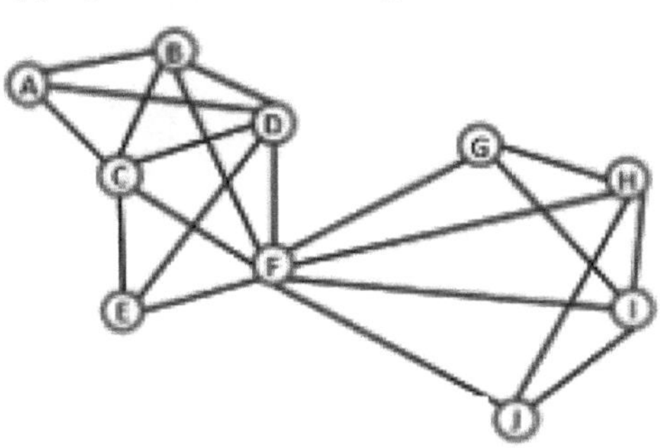

Q2. For the graph below:

a. Find nodes that are structurally equivalent nodes.
b. Calculate the modularity of the given network and detect the communities.

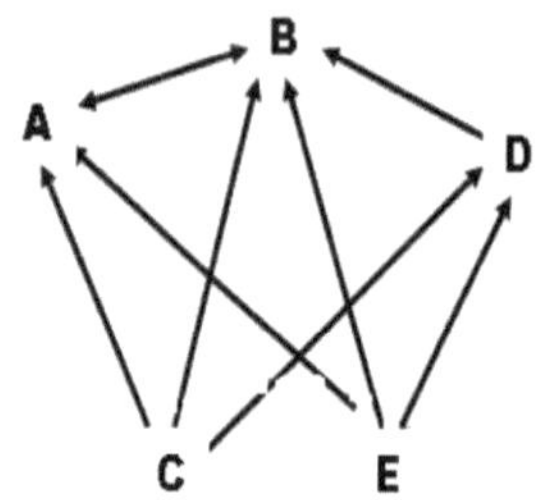

Q3. For the graph below:

a. Calculate similarity values between nodes v_2 and v_5.
b. Calculate 2-cliques, 2-clans and 2-clubs.

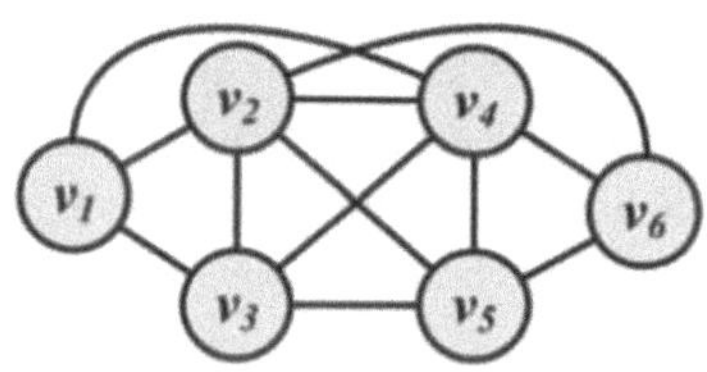

Q4. For the graph below:

a. Find out the maximum clique (show all the steps).
b. Find overlapping communities using the CPM for cliques of size 3.
c. Find out the structurally equivalent nodes.
d. Find the vertex similarity of the nodes (7,4).

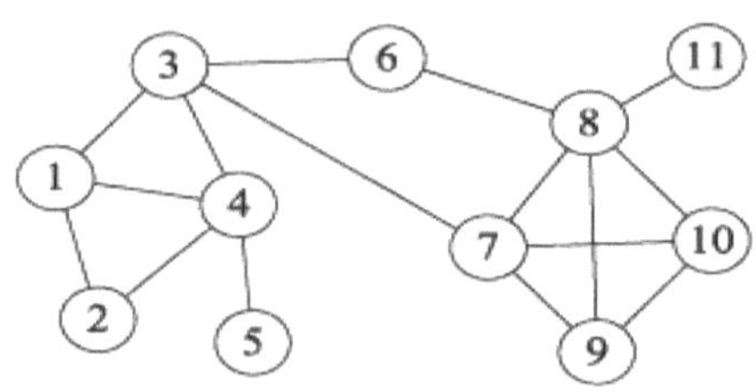

Q5. For the graph below, calculate the modularity of the given network and detect the communities.

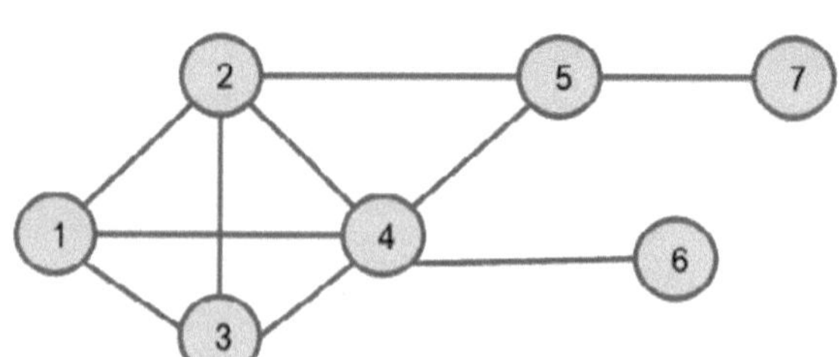

Q6. For the graph below, detect communities of the following network using the edge betweenness-based community structure detection method.

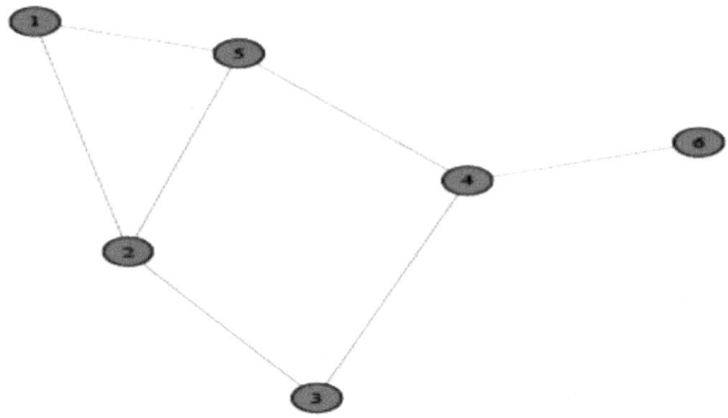

REFERENCES

1. Tang, L., & Liu, H. (2010). Community detection and mining in social media. *Synthesis Lectures on Data Mining and Knowledge Discovery*, 2(1), 1–137.
2. Ouyang, G. (2015). Social network community detection. (PhD diss. University of Connecticut).
3. Andalib, A. (2016, Nov 17). Community detection algorithms. https://www.slideshare.net/alirezaandalib77/community-detection-algorithms
4. Big data analysis. https://www.sciencedirect.com/topics/computer-science/big-data-analysis
5. Joshi, P. (2020, Apr 13). Getting started with community detection in graphs and networks. *Analytics Vidhya*. https://www.analyticsvidhya.com/blog/2020/04/community-detection-graphs-networks/
6. Luccio, F., & Sami, M. (1969). On the decomposition of networks in minimally interconnected subnetworks. *IEEE Transactions on Circuit Theory*, 16(2), 184–188.
7. Fasmer, E. E. (2015). Community detection in social networks (Master's thesis, The University of Bergen).
8. Borgatti, S. P., Everett, M. G., & Shirey, P. R. (1990). LS sets, Lambda sets and other cohesive subsets. *Social Networks*, 12(4), 337–357.
9. Sanei-Mehri, S. V., Das, A., Hashemi, H., & Tirthapura, S. (2021). Mining largest maximal quasi-cliques. *ACM Transactions on Knowledge Discovery from Data (TKDD)*, 15(5), 1–21.
10. Liu, H., Tang, L., & Agarwal, N. Community detection and behavior study for social behavior study for social computing (Presentation. Arizona State University).
11. Alasadi, A. H. H., Mohammed, H., & Alshemmary, E. N. (2013). Hybrid k-means clustering for color image segmentation. *Science and Engineering (IJCSE)*, 2(4), 17–26.
12. Wickelmaier, F. (2003). An introduction to MDS. *Sound Quality Research Unit, Aalborg University, Denmark*, 46(5), 1–26.
13. Varsha's Engineering Stuff. (2020, Mar 22). *Part I: Agglomerative hierarchical clustering algorithm, data mining, link single, complete, average* [YouTube video]. https://youtu.be/8rsJRbYnRR4
14. Varsha's Engineering Stuff. (2020, Mar 22). *Part II: Agglomerative hierarchical clustering algorithm, data mining, link single, complete, average* [YouTube video]. https://youtu.be/Y8gcg_PWAfc

15. Varsha's Engineering Stuff. (2020, Mar 22). *Part III: Agglomerative hierarchical clustering algorithm, data mining, link single, complete, average* [YouTube video]. https://youtu.be/hYnXR_OYBA0
16. Varsha's Engineering Stuff. (2020, Mar 22). *Part IV: Agglomerative hierarchical clustering algorithm, data mining, link single, complete, average* [YouTube video]. https://youtu.be/ZSc0VbPLVaM
17. Varsha's Engineering Stuff. (2020, Mar 23). Part I: Hierarchical Divisive Clustering Algorithm, Data Mining, Machine Learning, MST, example [YouTube video]. https://youtu.be/vQEXvV5W7s0
18. Varsha's Engineering Stuff. (2020, Mar 23). Part II: Hierarchical Divisive Clustering Algorithm, Data Mining, Machine Learning, MST, example [YouTube video]. https://youtu.be/4GbhWbMJLMY
19. ML: Types of linkages in clustering. https://www.geeksforgeeks.org/ml-types-of-linkages-in-clustering/
20. Hierarchical clustering: Foundational concepts and example of agglomerative clustering. https://www.dexlabanalytics.com/blog/hierarchical-clustering-foundational-concepts-and-example-of-agglomerative-clustering
21. Poddar, R., & Shrivastava, M. Complex network theory. https://cse.iitkgp.ac.in/~animeshm/scribe.pdf
22. Bhatia, A. (2019, May 19). *Girvan Newman BDA*. [YouTube video]. https://www.youtube.com/watch?v=LtQoPEKKRYM&t=494s

8 Ego Networks

8.1 OVERVIEW OF EGO NETWORKS

Ego networks consist of a focal node ('ego') and the nodes to whom the ego is directly connected to (these are called 'alters') plus the ties, if any, among the alters.

Of course, each alter in an ego network has their own ego network, and all ego networks interlock to form the human social network.

Social Relations: Egos and alters are tied to each other by social relations. There are various types of social relations, as shown in Figure 8.2.

Data Collection for Ego Networks: Data collection for ego networks can be done with two basic strategies: person-based and relation-based. The person-based strategy uses a set of open-ended questionnaire items known as a name generator. Typically, it is unlimited in scope: the respondent may name anyone from any sphere of life: neighbours, kin, friends, co-workers etc. After obtaining a large list of names, the interviewer typically goes over each name, asking the respondent about the nature of their relationship with that person (what social relation) and asking about attributes of that person (sex, race, income etc.). When possible, this is then followed by asking the respondent to indicate relationships among the alters [1].

The relation-based strategy starts with a relation of interest, such as emotional support, and then asks all the people that the respondent has this particular relationship with. Sometimes this is within the context of a circumscribed group, such as 'which of your fellow employees do you obtain emotion support from?' If appropriate this is then followed by attributes of the tie, such as duration, intensity, frequency, strength and so on. If the relation is 'friendship', it might ask about strength and duration, but not frequency. If the relation is 'interacts with', it might ask about frequency and tenure of the relationship. Attributes of the alters can also be asked.

It should be obvious that random sampling of egos from a large population will not form a network: it is likely that no ego mentions an alter that is another ego or is

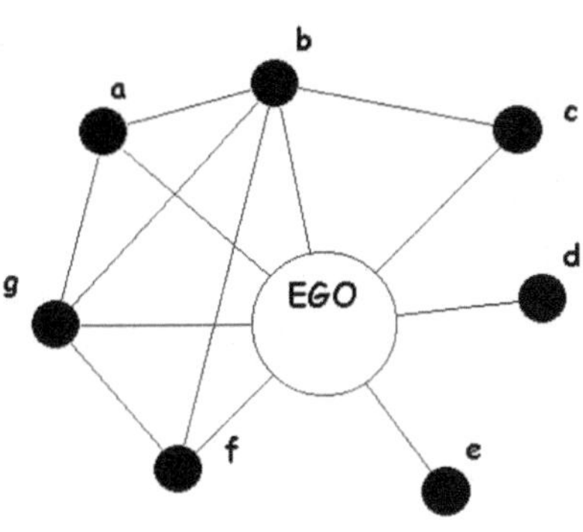

FIGURE 8.1 Ego network

DOI: 10.1201/9781003088066-8

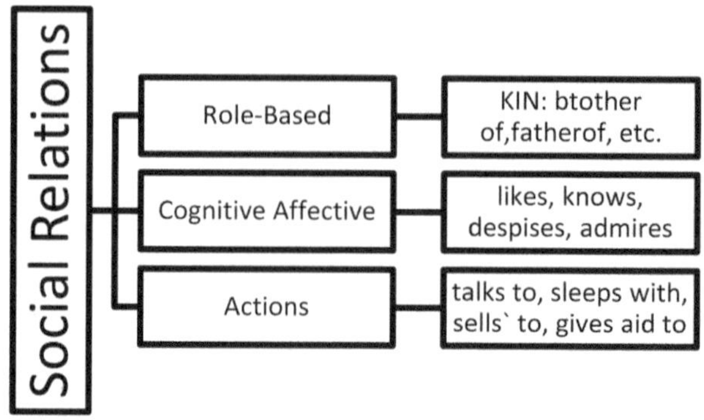

FIGURE 8.2 Social relations

an alter of another ego. Hence, certain network measures cannot be obtained from ego network data, such as the number of links that separate people.

Functions Served by Ego Networks: The main functions that are served by ego networks are social support, emotional and material aid, companionship, information, sense-making, how to interpret the world, social control, ensuring that egos behave according to norms, access to resources, entrepreneurs drawing on their contacts to get the clients and employees and consultants they need, behavioural models, you tend to talk like the people you talk to and you become aware of choices – from clothing options to brands of cars to ways of talking and behaving – from the people in your ego network [2].

8.2 CHARACTERISTICS OF EGO NETWORKS

A standing hypothesis about ego networks is that strong ties are homophilous. That is, people have the strongest ties with people who are similar to themselves on key attributes, such as social class, age, sex, race, political views etc. Another standing

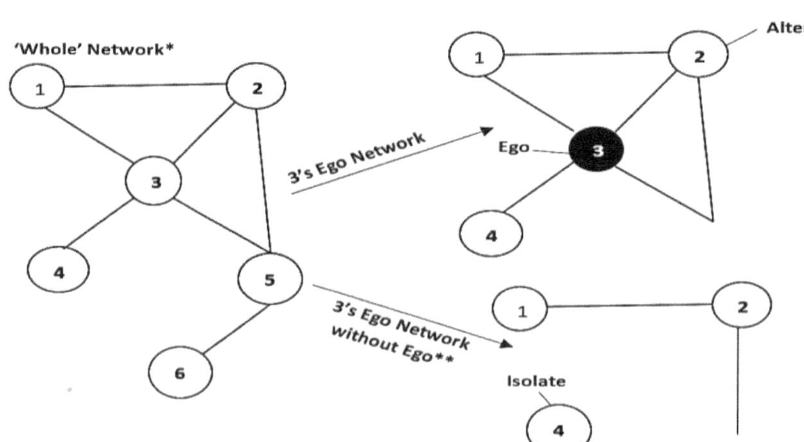

FIGURE 8.3 Ego networks and whole networks

hypothesis or guiding principle is that people with heterogeneous networks are 'better off'. The greater the diversity of their network [3], the more chance that someone in the network has something the ego needs. This is particularly relevant for entrepreneurs.

According to Granovetter [4], the stronger the tie between the ego and two of her alters, the greater the likelihood that the alters enjoy at least a weak tie. Granovetter also believes that weak ties provide people with access to novel information, whereas your strong ties tend to know the same things you do. Hence, the denser the ties in an ego network, the stronger the ties, and the more insular the ego network and also the more homogeneous.

8.3 EGO NETWORK MEASURES

8.3.1 Ego Network Density

Ego network density can be measured like normal density but we have to ignore ego–alter ties. The proportion of ties between alters is compared to the number possible.

For example:

$$\text{Total ties between alters} = 1$$

$$\text{Possible} = \frac{n*(n-1)}{2} = 4*3/2 = 6$$

$$\text{Density} = 1/6$$

8.3.2 Structural Holes

If we compare two nodes, node A is more likely to get novel information than node B, even though they have the same number of links. This is so because nodes connected to B are also highly connected between each other. Therefore, any information that any of them could get from B, it could easily get from other nodes as well. Furthermore, the information that B gets from different connections is likely to be overlapping, so connections involving node B are said to be redundant.

On contrary, the position of node A makes it serve as a bridge or a 'broker' between three different clusters. Thus, node A is likely to receive some non-redundant information from its contacts. The term 'structural hole' is used for the separation between

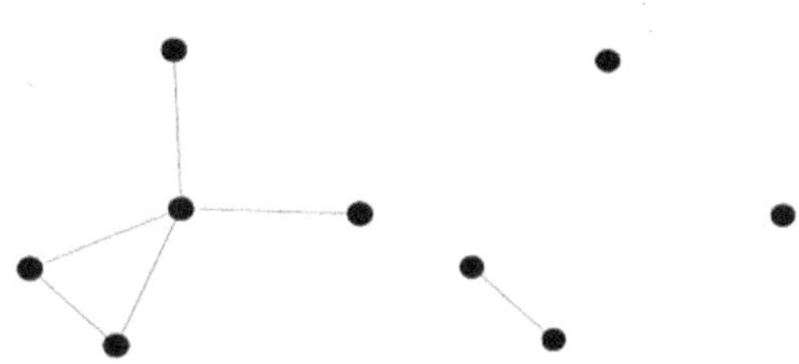

FIGURE 8.4 Example of ego network density

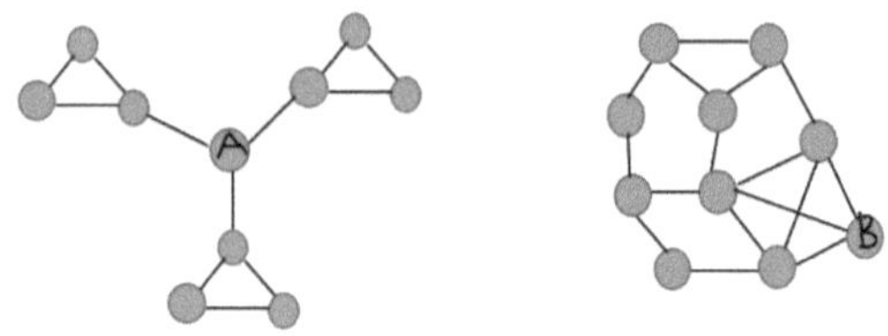

FIGURE 8.5 Structural holes

non-redundant contacts. As a result of the hole between two contacts, they provide network benefits to the third party (to node A).

Structural holes are 'disconnections or non-equivalencies between players' and hence 'entrepreneurial opportunities for information access, timing, referrals, and control' [5]. There is a structural hole between two players if there is a potential for beneficial information flow between them. The word disconnections in the above definition refer to the absence of a tie or path through which the information can flow.

A network rich in structural holes thus contains many exploitable brokering opportunities: 'The structural hole is an opportunity to broker the flow of information between people and to control the form of projects that bring together people from opposite sides of the hole'. The structural entrepreneur recognizes these opportunities and places themselves in the hole by initiating ties with both players (a person who generates profit from being between others).

Structural Hole Measures

Bridge counts: a bridge count is a simple and intuitive measure of structural holes in a network. A bridge is defined as a relation between two individuals if there is no indirect connection between them through mutual contacts.

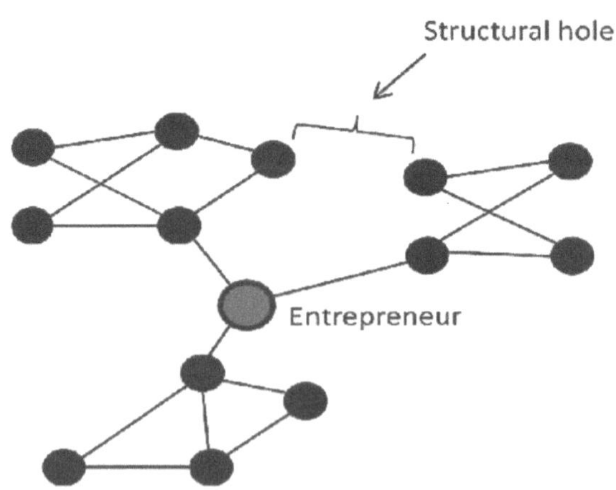

FIGURE 8.6 Structural hole example

FIGURE 8.7 Effective size example

Effective size: an ego's number of alters minus the average number of ties that each alter has to other alters. The effective size is a positive function of network size, and a negative function of the number of ties among alters.

Burt's Formula (Burt's Constraint)

Burt's formula is commonly used as a measure of structural holes (accurately speaking, the lack of it, because the larger the constraint value, the less structural opportunities a node has for bridging structural holes). It is suggested by Burt as a measurement for the concept of 'bridging social capital'. Constraint measures how much the ego's friends are also connected among them.

Nodes with lower constraint values are not as constrained by her friends (i.e., you get to access new information outside of a highly cohesive friends' pool). Nodes with higher constraint values are constrained by her friends (i.e., you don't get as much exposure to information outside of what's already circulating in a highly cohesive friend's pool). People who act as information brokers have lower constraint values.

Think about the kind of people who seem to know everyone – they usually have lower constraint scores. Those who are less socially active and prefer to stay within a closed social circle, such as within a classmate network or a department network, tend to have higher constraint scores.

Constraint is not the only measure of structural holes, although it is the most often used one (according to my observation). Burt also proposed two other measures in parallel to constraint: bridge counts and effective size.

Burt introduced the measure of a network's redundancy. He aims to estimate to what extent contact j is redundant with other contacts of node i. Redundancy is understood as an investment of time and energy in a relationship with another node q, with whom node j is strongly connected

$$\text{Redundancy} = p_{iq} m_{qj} \tag{8.1}$$

where p_{iq} is proportion of i's energy invested in relationship with q, and m_{qj} is calculated as j's interaction with q divided by j's strongest relationship with anyone.

The redundancy in network is calculated by summing up this product across all nodes q. One minus this expression expresses the non-redundant portion of relationship. The effective size of i's network is defined as a sum of the j's non-redundant contacts

$$\text{Effective size of i's network} = \sum_j \left[1 - \sum_q p_{iq} m_{qj} \right] q \neq i, j, \tag{8.2}$$

The more each node is disconnected from other primary contacts, the higher the effective size would be. This indicator varies from 1 (network only provides a single link) to the total number of links N (each contact is non-redundant).

Borgatti's reformulation of Burt's formula: Borgatti developed a simplified formula to calculate the effective size for unweighted networks

$$\text{Redundancy} = \frac{2t}{n} \tag{8.3}$$

where t is the number of the total ties in the egocentric network (excluding those ties to the ego) and n is the number of total nodes in the egocentric network (excluding the ego).

This formula can be modified to calculate the effective size of the ego's network

$$\text{Effective size of ego's network} = n - \frac{2t}{n} \tag{8.4}$$

Burt's constraint: the network constraint of a network is a sum of each connection's constraints c_{ij}

$$c_{ij} = \left(p_{ij} + \sum_q p_{iq} p_{qj} \right)^2, \ \mathrm{i} \neq q \neq j \tag{8.5}$$

This indicator measures the extent to which time and energy is concentrated within a single cluster. It consists of two components:

Direct, when a contact consumes a large proportion of a network's time and energy (P_{ij} – direct influence). Burt assumes that an actor distributes their time equally over their contacts: if i is connected to j, then, where $p_{ij} = \frac{1}{d_{ij}}$, where d_{ij} is actor i's degree – that is, the number of ties of i. If i and j are not connected, $p_{ij} = 0$.

Indirect, when a contact controls other individuals who consume a large proportion of a network's time and energy [6]. (sum $(P_{iq} * P_{qj})$ – indirect influence).

Network constraint measures the extent to which the manager's network of colleagues is like a straitjacket around the manager, limiting their vision of alternative ideas and sources of support. It depends on three network characteristics: size, density and hierarchy. Constraint on an individual would be generally higher in the case of a small network (they have just a few contacts), and if contacts are highly connected between each other (either directly as in a dense network, or indirectly, through the mutual central contact as in a hierarchical network).

Burt's constraint explanation: P_{ij} is the proportion of edge weights from i to j, where i is the ego and j is the alter. You can summarize this part as: How much resources/energy/communication effort has the ego spent on the alter through direct paths?

Say that the ego talked with five people today and one of them is the alter, then the direct path has an edge weight of 1. Thus $P_{ij} = 1/5$. 1 *out of* 5 units of resources are spent directly from the ego to the alter. $P_{iq} * P_{qj}$ is measuring the amount of resources/energy/communication effort the ego and the alter can indirectly share [7].

Assume there exists an indirect path

$$\text{node } i \text{ — node } q \text{ — node } j$$

Then, the product of proportion of the edge weights between i to q, and q to j will be the total amount of indirect influence from ego to alter. It is quite intuitive. Assume that node i spent another 1/5 of her energy today communicating with node q, and node q also spent 1/3 energy talking to node j. Then it is like the ego i indirectly spent $1/5 * 1/3$ of her energy on alter j. Then the rest is easy. Sum up this value for each alter in the ego's local network, and calculate the squared value.

The constraint measure – three example networks:

Burt's constraint:

$$c_{ij} = \left(p_{ij} + \sum_q p_{iq} p_{qj} \right)^2, \ i \neq q \neq j$$

where,

$$p_{ij} = \frac{1}{d_{ij}} \tag{8.6}$$

$$d_A = 4;\ d_B = 1;\ d_C = 1;\ d_D = 1;\ d_E = 1;\ P_A = 1/4;\ P_B = 1;\ P_C = 1;\ P_D = 1;\ P_E = 1$$

$$C_A = C_{AB} + C_{AC} + C_{AD} + C_{AE};\ C_A = \left(\frac{1}{4} + 0\right)^2 + \left(\frac{1}{4} + 0\right)^2 + \left(\frac{1}{4} + 0\right)^2 + \left(\frac{1}{4} + 0\right)^2;$$

$$C_A = 4 * \left(\frac{1}{4} + 0\right)^2 = 4 * \frac{1}{16} = \frac{1}{4}$$

$$C_B = C_{BA} + C_{BC} + C_{BD} + C_{BE};\ C_B = \left(\frac{1}{1} + 0\right)^2 + 0 + 0 + 0 = 1$$

$$C_C = C_{CA} + C_{CB} + C_{CD} + C_{CE};\ C_C = \left(\frac{1}{1} + 0\right)^2 + 0 + 0 + 0 = 1$$

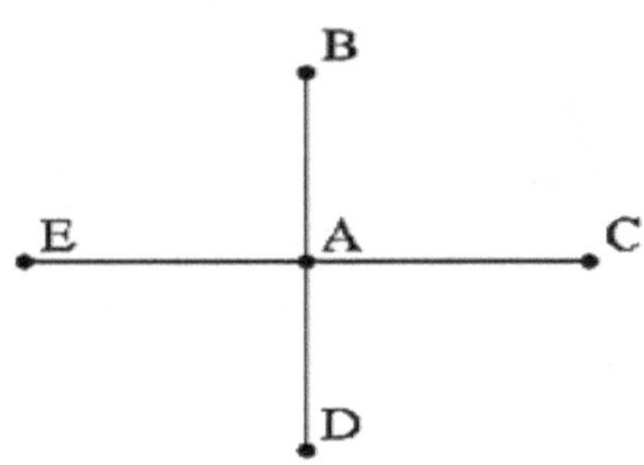

FIGURE 8.8 Structural hole example 1

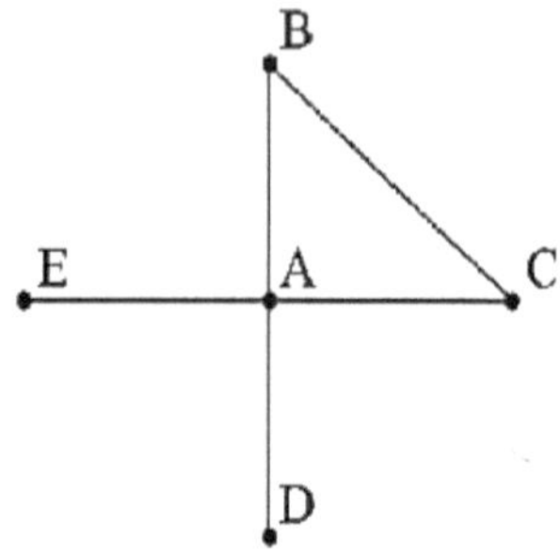

FIGURE 8.9 Structural hole example 2

$$C_D = C_{DA} + C_{DB} + C_{DC} + C_{DE};\ C_D = \left(\frac{1}{1} + 0\right)^2 + 0 + 0 + 0 = 1$$

$$C_E = C_{EA} + C_{EB} + C_{EC} + C_{ED};\ C_E = \left(\frac{1}{1} + 0\right)^2 + 0 + 0 + 0 = 1$$

$$d_A = 4;\ d_B = 2;\ d_C = 2;\ d_D = 1;\ d_E = 1;\ P_A = 1/4;\ P_B = 1/2;\ P_C = 1/2;\ P_D = 1;\ P_E = 1$$

$$C_A = C_{AB} + C_{AD} + C_{AC} + C_{AE}$$

$$C_A = \left(\frac{1}{4} + \frac{1}{4}.\frac{1}{2}\right)^2 + \left(\frac{1}{4} + 0\right)^2 + \left(\frac{1}{4} + \frac{1}{4}.\frac{1}{2}\right)^2 + \left(\frac{1}{4} + 0\right)^2$$

$$C_A = 2 * \left(\frac{1}{4}\right)^2 + 2 * \left(\frac{1}{4} + \frac{1}{8}\right)^2 = 2 * \frac{1}{16} + 2 * \left(\frac{3}{8}\right)^2 = \frac{1}{8} + \frac{9}{32} = \frac{13}{32}$$

$$C_B = C_{BA} + C_{BC} + C_{BD} + C_{BE}$$

$$C_B = \left(\frac{1}{2} + \frac{1}{2} \times \frac{1}{4}\right)^2 + \left(\frac{1}{2} + \frac{1}{2} \times \frac{1}{2}\right)^2 + 0 + 0 = \left(\frac{1}{2} + \frac{1}{8}\right)^2 + \left(\frac{1}{2} + \frac{1}{4}\right)^2$$

$$C_B = \frac{25}{64} + \frac{9}{16} = \frac{61}{64}$$

$$C_C = C_{CA} + C_{CB} + C_{CD} + C_{CE};\ C_C = \left(\frac{1}{2} + \frac{1}{2} \times \frac{1}{4}\right)^2 + \left(\frac{1}{2} + \frac{1}{2} \times \frac{1}{2}\right)^2 + 0 + 0$$

$$C_C = \left(\frac{1}{2} + \frac{1}{8}\right)^2 + \left(\frac{1}{2} + \frac{1}{4}\right)^2;\ C_B = \frac{25}{64} + \frac{9}{16} = \frac{61}{64}$$

$$C_D = C_{DA} + C_{DB} + C_{DC} + C_{DE};\ C_D = \left(\frac{1}{1} + 0\right)^2 + 0 + 0 + 0 = 1$$

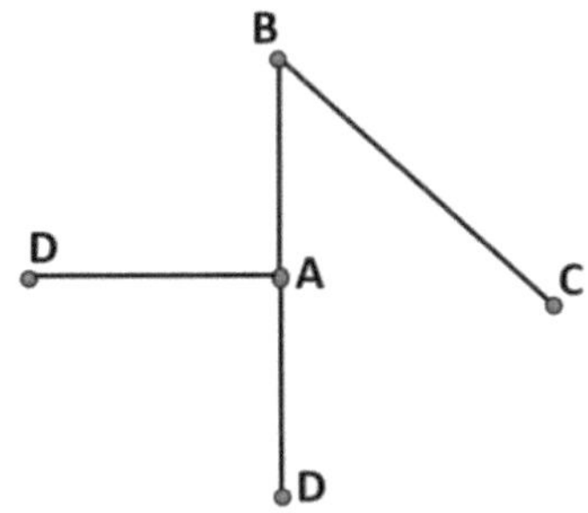

FIGURE 8.10 Structural hole example 3

$$C_E = C_{EA} + C_{EB} + C_{EC} + C_{ED};\ C_E = \left(\frac{1}{1}+0\right)^2 + 0 + 0 + 0 = 1$$

$$d_A = 3;\ d_B = 2;\ d_C = 1;\ d_D = 1;\ d_E = 1;\ P_A = 1/3;\ P_B = 1/2;\ P_C = 1;\ P_D = 1;\ P_E = 1$$

$$C_A = C_{AB} + C_{AC} + C_{AD} + C_{AE};\ C_A = \left(\frac{1}{3}+0\right)^2 + 0 + \left(\frac{1}{3}+0\right)^2 + \left(\frac{1}{3}+0\right)^2 = 3*\left(\frac{1}{3}\right)^2 = \frac{1}{3}$$

$$C_B = C_{BA} + C_{BC} + C_{BD} + C_{BE};\ C_B = \left(\frac{1}{2}+0\right)^2 + \left(\frac{1}{2}+0\right)^2 + 0 + 0 = \frac{1}{2}$$

$$C_C = C_{CA} + C_{CB} + C_{CD} + C_{CE};\ C_C = 0 + \left(\frac{1}{1}+0\right)^2 + 0 + 0 = \frac{1}{2}$$

$$C_D = C_{DA} + C_{DB} + C_{DC} + C_{DE};\ C_D = \left(\frac{1}{1}+0\right)^2 + 0 + 0 + 0 = 1$$

$$C_E = C_{EA} + C_{EB} + C_{EC} + C_{ED};\ C_E = \left(\frac{1}{1}+0\right)^2 + 0 + 0 + 0 = 1$$

8.3.3 Brokerage

A broker connects clusters (communities). Brokerage is a process in which individuals called brokers act as intermediaries between individuals or groups who do not have direct access to each other. The broker provides a link between those segmented or isolated groups or individuals so that access to goods, services or information is enabled. Brokers possess specialized knowledge or resources that enable them to act more effectively than those individuals or groups could themselves.

Brokerage Model

Among the network analysis techniques, the brokerage model is one of the ego networks that consist of a focal actor as a set of alters that have ties with the ego (which is particularly located in the center as a target) as well as measurements on the ties

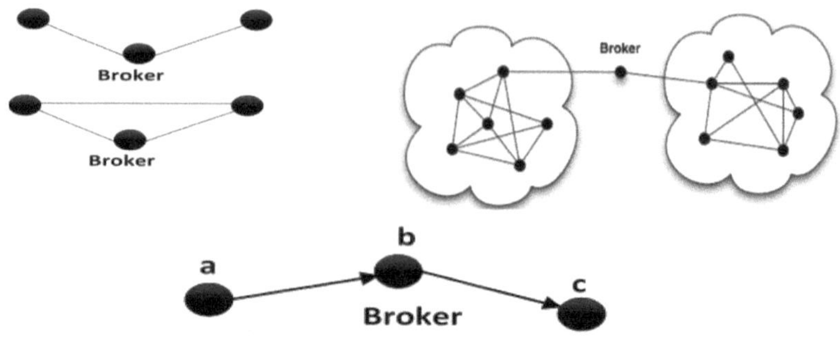

FIGURE 8.11 Broker example

among these alters [8]. When the actor – as the 'ego' – is located in the centre, other alters are related around this ego.

In the brokerage model, the ego is considered the broker, and the ego is the internal source related to other internal or external sources in the view of open innovation [3]. To sum up, through the group relationships analysis, the brokerage analysis aims to find such relationships between groups or fields based on the broker [9]. Five types of brokers are suggested and widely accepted as the representative brokerage model [10].

Brokerage Roles

Coordinator: all nodes belong to the same group.

Consultant: based on the broker in group A, both the giver and the recipient belong to the different group B.

Gatekeeper: based on the broker in the group A, the giver belongs to a different group B, while the recipient belongs to the same group A.

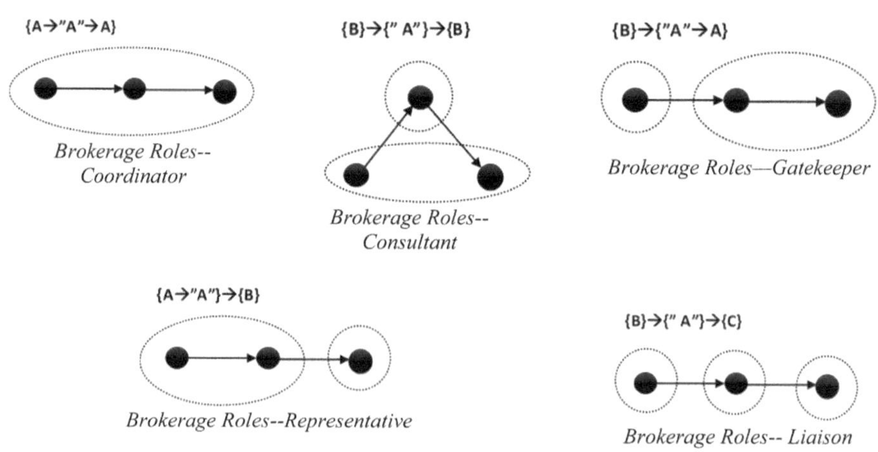

FIGURE 8.12 Illustration of each brokerage type

Representative: based on the broker in group A, the recipient belongs to a different group B, while the giver belongs to the same group A.

Liaison: all nodes belong to different groups.

Note: 'A' in the middle is the broker. We can count how often a node enacts each kind of brokerage role. So far, we have identified brokerage with a particular network shape. But brokerage can also occur when the brokered are already connected – a catalyst (agent) to do something. Marriage and real estate brokers both exist to create a tie of some kind.

8.4 PROBLEMS FOR SELF-ASSESSMENT

Q1. What is the density of the graph below?

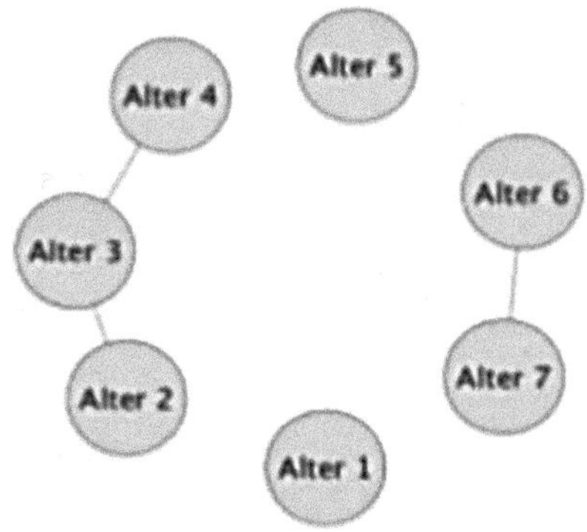

Q2. Measure the structural hole in the following network using the constraint formula (Burt's constraint).

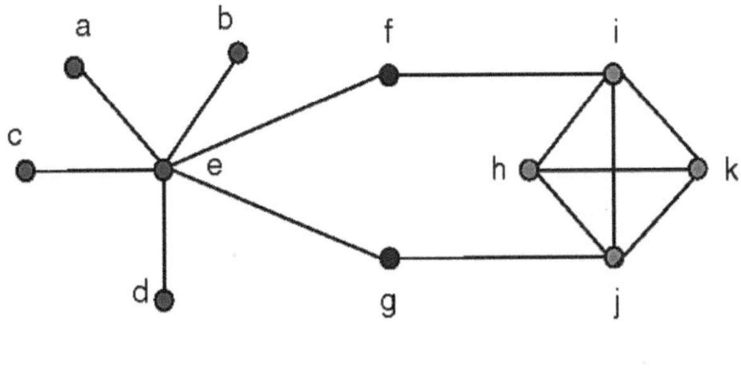

REFERENCES

1. Borgatti, S. Ego networks. (1998). *Social Network Analysis Instructional Web Site.* http://www.analytictech.com/networks/egonet.htm
2. Hanneman, R., & Riddle, M. (2005). *Introduction to Social Network Methods: Ego Networks.* University of California, Riverside. https://faculty.ucr.edu/~hanneman/nettext/C9_Ego_networks.html
3. Suh, Y., & Kim, M. S. (2015). Dynamic change of manufacturing and service industries network in mobile ecosystems: The case of Korea. *Telematics and Informatics*, 32(4), 613–628.

4. Granovetter, M. S. (1973). The strength of weak ties. *American Journal of Sociology*, 78(6), 1360–1380.
5. Wauchope, R. D., Buttler, T. M., Hornsby, A. G., Augustijn-Beckers, P. W. M., & Burt, J. P. (1992). The SCS/ARS/CES pesticide properties database for environmental decision-making. *Reviews of Environmental Contamination and Toxicology*, 123, 1–155.
6. Wikipedia. Structural holes. https://en.wikipedia.org/wiki/Structural_holes
7. Ren, R. (2018, Mar 1). Hand calculation of Burt's constraint in networks. https://ruqinren.wordpress.com/2018/03/01/hand-calculation-of-burts-constraint-in-networks/
8. Wasserman, S., & Faust, K. (1994). Social network analysis: Methods and applications.
9. Jeon, J., & Suh, Y. (2019). Multiple patent network analysis for identifying safety technology convergence. *Data Technologies and Applications*, 53(2), 269–285.
10. Gould, R. V., & Fernandez, R. M. (1989). Structures of mediation: A formal approach to brokerage in transaction networks. *Sociological Methodology*, 19, 89–126.

9 Network Cohesion

9.1 OVERVIEW OF NETWORK COHESION

In social network analysis, the term *network cohesion* refers to a measure of the connectedness and togetherness among actors within a network. A social network can be defined as a network formed by a set of interacting social entities (actors) and the linkages (relations or edges) among them. The index of network cohesion is a single value that captures the togetherness of the group. Network cohesion can be measured in a variety of different ways, most of which are based on the dyadic cohesion or triadic cohesion [1].

Until now, we treated networks largely as static structures – we take a snapshot of the nodes and edges at a particular moment in time, and then ask about paths, components, distances and so forth. While this style of analysis forms the basic foundation for thinking about networks, it is also useful to think about how a network evolves over time [2]. In particular, what are the mechanisms by which nodes arrive and depart, and by which edges form and vanish? If two people in a social network have a friend in common, then there is an increased likelihood that they will become friends themselves at some point in the future. We refer to this principle as triadic closure.

9.2 TRIADIC CLOSURE

If nodes B and C have a friend A in common, then the formation of an edge between B and C produces a situation in which all three nodes A, B and C have edges connecting each other – a structure we refer to as a *triangle* in the network. The term *triadic closure* comes from the fact that the B–C edge has the effect of 'closing' the third side of this triangle. The formation of the edge between B and C illustrates the effects of triadic closure, since they have a common neighbour, *A*.

If we watch a network for a longer span of time, we can see multiple edges forming – some form through triadic closure while others (such as the D–G edge) form

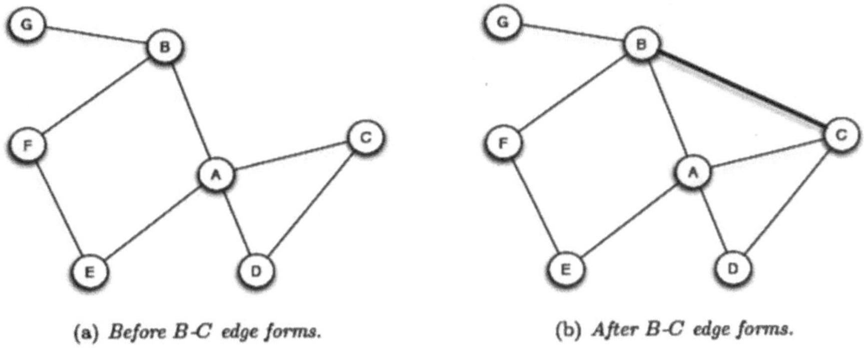

FIGURE 9.1 Example effect of triadic closure [1]

DOI: 10.1201/9781003088066-9

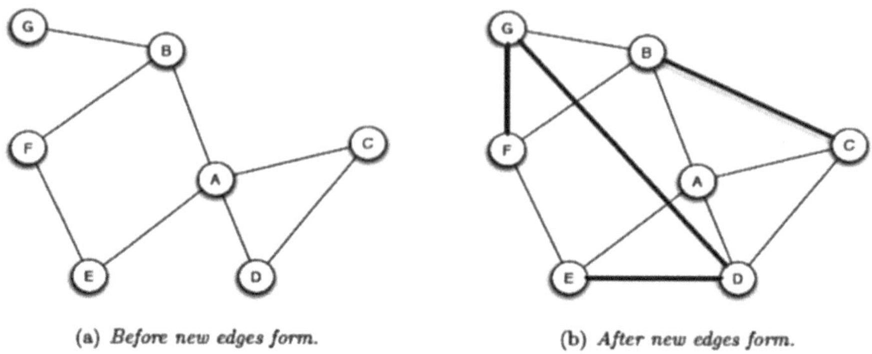

FIGURE 9.2 Example effect of triadic closure after a long time span [1]

even though the two endpoints have no neighbours in common. Thus, the Principle of Triadic Closure: if two people have a friend in common, then there is increased likelihood that they will become friends in the future –G B F E A C D G B F E A C D

Measuring triadic closure: the clustering coefficient of a node A is the probability that two random friends of A are connected to the fraction of pairs of A's friends that are connected: G B F E A C D.

9.3 EMBEDDEDNESS

Let's start with node A. Node A's set of network neighbours has been subject to considerable triadic closure; A has a high clustering coefficient. (Recall that the clustering coefficient is the fraction of pairs of neighbours who are themselves neighbours.)

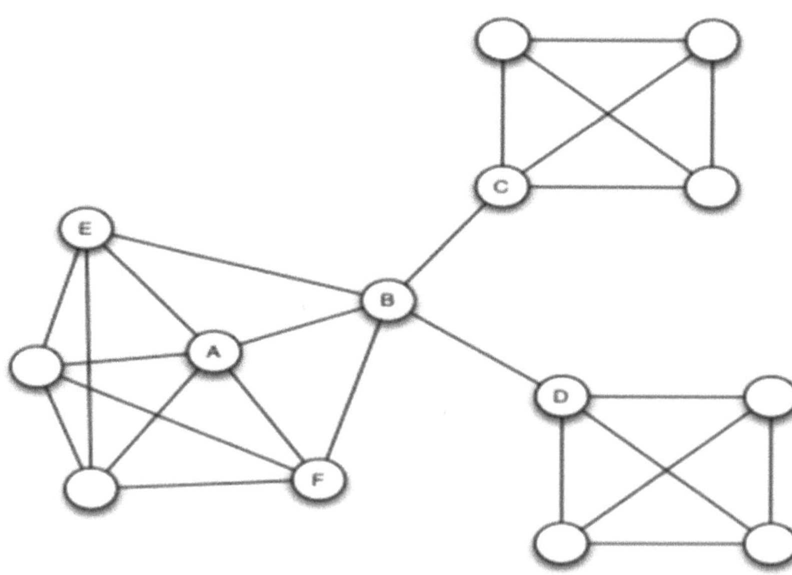

FIGURE 9.3 Example of embeddedness

To talk about the structure around A, it is useful to introduce an additional definition. We define the embeddedness of an edge in a network to be the number of common neighbours that the two endpoints have [3].

Thus, for example, the A–B edge has an embeddedness of two, since A and B have two common neighbours, E and F.

9.4 DENSITY

A network's *density* is the ratio of the number of edges in the network over the total number of possible edges between all pairs of nodes.

$$\text{Density} = n\frac{n-1}{2} \tag{9.1}$$

where n is the number of vertices for an undirected graph.

It is a common measure of how well connected a network is (in other words, how closely knit it is) – a perfectly connected network is called a *clique* and has a *density* = 1.

In Figure 9.4(a), density = 5/6 = 0.83 (i.e., it is a fairly *dense* network; the opposite would be a *sparse* network).

A directed graph will have half the density of its undirected equivalent, because there are twice as many possible edges, i.e., $n(n-1)$.

In Figure 9.4(b), density = 5/12 = 0.42.

Density is useful in comparing networks against each other, or in doing the same for different regions within a single network.

9.5 DYADIC RELATION

The smallest social structure in which an individual can be embedded is a dyad (that is, a pair of actors). For binary ties (present or absent), there are two possibilities for each pair in the population – either they have a tie or they don't. We can characterize the whole population in terms of the prevalence of these dyadic 'structures'. This is what the density measure does.

If we are considering a directed relation (A might like B, but B might not like A), there are three kinds of dyads (no tie, one likes the other but not vice versa or both like the other). The extent to which a population is characterized by 'reciprocated' ties (those where each directs a tie to the other) may tell us about the degree of cohesion, trust and social capital that is present.

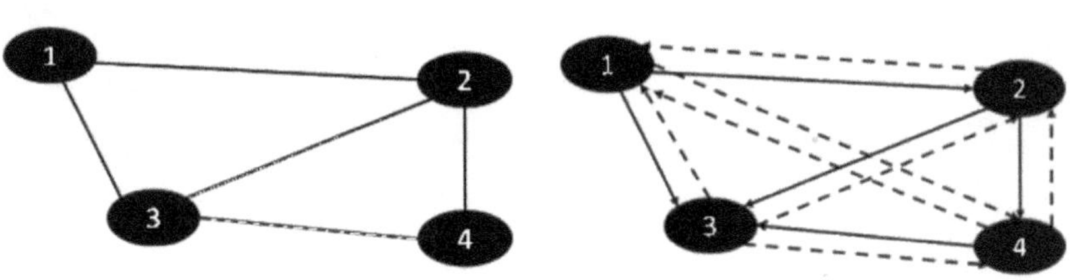

FIGURE 9.4 Example of density

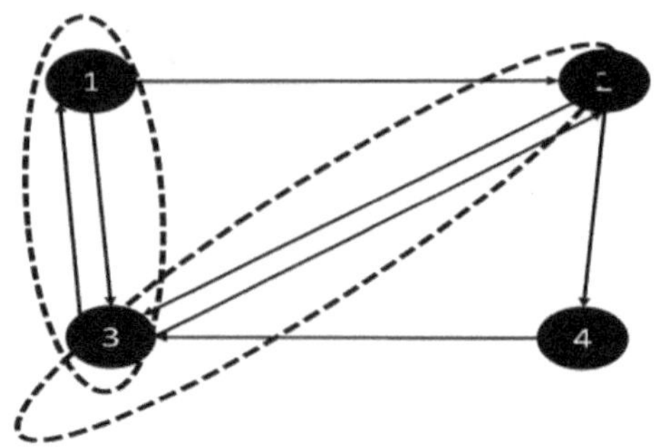

FIGURE 9.5 Example of reciprocity

9.6 RECIPROCITY

In network science, reciprocity is a measure of the likelihood of vertices in a directed network to be mutually linked. The ratio of the number of relations which are reciprocated (i.e., there is an edge in both directions) over the total number of relations in the network where two vertices are said to be related if there is at least one edge between them.

In Figure 9.5, this would be $2/5 = 0.4$ (whether this is considered high or low depends on the context). This is a useful indicator of the degree of mutuality and reciprocal exchange in a network, which relate to social cohesion and it only makes sense in directed graphs.

Reciprocity can be seen as direct and indirect reciprocity:

1. Direct reciprocity means that A helps B and B helps A.
2. Indirect reciprocity comes in two types:
   - Upstream reciprocity (Figure 9.6[a]) is based on a recent positive experience. A person who has been at the receiving end of a donation may feel motivated to donate in turn. Individual B, who has just received help from A, goes on to help C.
   - Downstream reciprocity (Figure 9.6[b]) is built on reputation. Individual A has helped B and therefore receives help from C

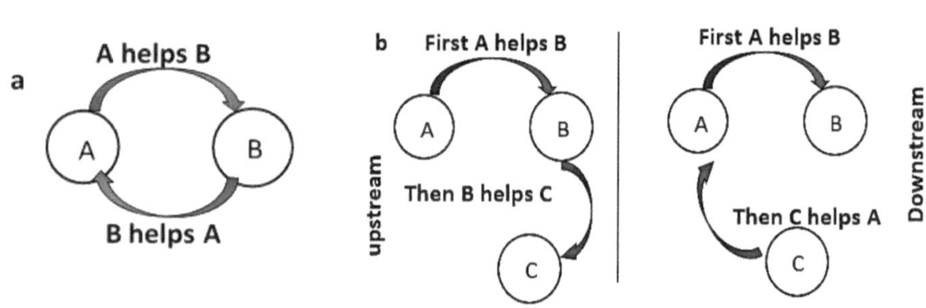

FIGURE 9.6 Reciprocity types

With directed data, there are four possible dyadic relationships:

1. A and B are not connected.
2. A sends to B.
3. B sends to A.
4. A and B send to each other.

A common interest in looking at directed dyadic relationships is the extent to which ties are reciprocated. Some theorists feel that there is an equilibrium tendency towards dyadic relationships to be either null or reciprocated, and that asymmetric ties may be unstable. A network that has a predominance of null or reciprocated ties over asymmetric connections may be a more 'equal' or 'stable' network than one with a predominance of asymmetric connections (which might be more of a hierarchy). There are (at least) two different approaches to indexing the degree of reciprocity in a population. Consider the very simple network shown in Figure 9.7.

- Actors A and B have reciprocated ties.
- Actors B and C have a non-reciprocated tie.
- Actors A and C have no tie.

Reciprocity example: What is the prevalence of reciprocity in this network?

One approach is to focus on the dyads, and ask what proportion of pairs has a reciprocated tie between them. This would yield one such tie for three possible pairs (AB, AC and BC), or a reciprocity rate of 0.333. More commonly, analysts are concerned with the ratio of the number of pairs with a reciprocated tie relative to the number of pairs with any tie.

In large populations, usually most actors have no direct ties to most other actors, and it may be more sensible to focus on the degree of reciprocity among pairs that have any ties. In our simple example, this would yield one reciprocated pair divided by two tied pairs, or a reciprocity rate of 0.500. The method just described is called the dyad method.

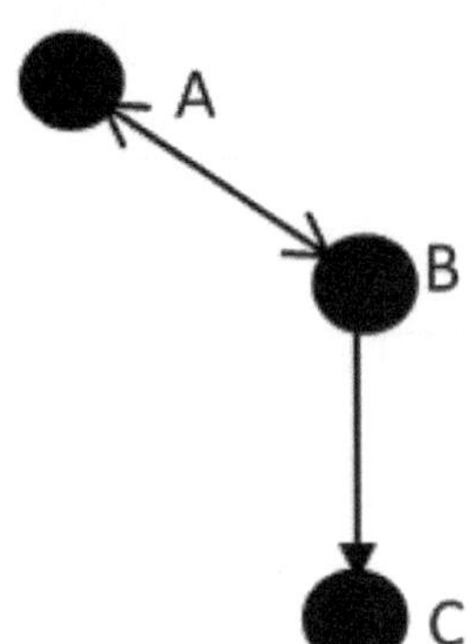

FIGURE 9.7 Example graph for reciprocity

Rather than focusing on actors, we could focus on relations. We could ask what percentage of all possible ties (or 'arcs') of the directed graph is part of reciprocated structures?

Here, two such ties (A to B and B to A) are a reciprocated structure among the six possible ties (AB, BA, AC, CA, BC and CA) or reciprocity of 0.333. (% of pairs that have reciprocated ties/all possible pairs AB of {AB, AC, BC} = 0.33

Analysts usually focus, instead, on the number of ties that are involved in reciprocal relations relative to the total number of actual ties (not possible ties). Here, this definition would give us 2/3 or 0.667. This approach is called the arc method.

Thus, to answer 'What is the reciprocity in this network?', there can be three most relevant answers:

Answer 1: % of pairs that have reciprocated ties/all possible pairs AB of {AB, AC, BC} = 0.33

Answer 2: % of pairs that have reciprocated ties/existing pairs AB of {AB, BC} = 0.5

Answer 3: % directed ties/all directed ties {AB, BA} of {AB, BA, AC, CA, BC, CA} = 0.33

Directed networks like the follow network of Twitter are not inherently reciprocal, which means edges existing from A to B do not imply the existence of a relationship from B to A. This type of relationship can also be seen in citation networks. However, friendship networks of Facebook or co-authorship networks of scientists are reciprocal relationships.

9.7 HOMOPHILY

Homophily is the tendency to relate to people with similar characteristics (status, beliefs etc.) It leads to the formation of homogeneous groups (*clusters*) where forming relations is easier. Extreme homogenization can act counter to innovation and idea generation (*heterophily* is thus desirable in some contexts). Homophilous ties can be strong or weak.

Let's suppose that we have the friendship network of an elementary-school classroom, and we suspect that it exhibits homophily by gender: boys tend to be friends with boys, and girls tend to be friends with girls.

For example, the graph shows the friendship network of a (small) hypothetical classroom in which the three shaded nodes are girls and the six unshaded nodes are boys.

If there were no cross-gender edges at all, then the question of homophily would be easy to resolve: it would be present in an extreme sense.

Principle of homophily: people tend to be similar to their friends. Think about it:

- Does triadic closure encourage or discourage homophily?
- How can we discern (recognize) if a network exhibits homophily with respect to a given property (e.g., Do people tend to be friends with others of the same race in the network?)
- Do people tend to be friends with others of the same social status in the network?

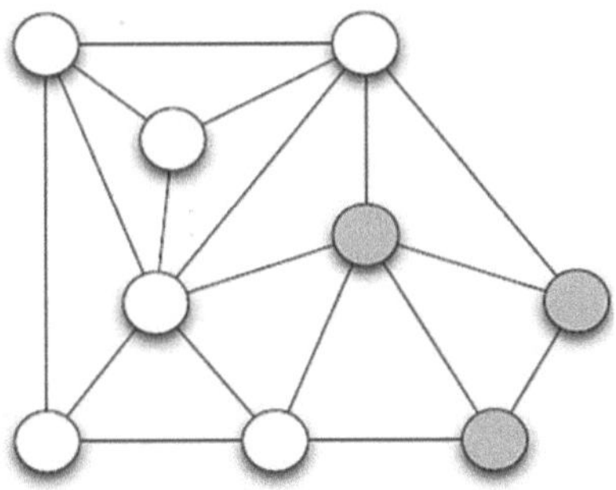

FIGURE 9.8 Example graph for homophily

Consider a property that has two values. Each person has exactly one of these two values: male or female. Suppose that a fraction p of people in the network is male, while the fraction $q = 1 - p$ is female. If nodes were randomly assigned 'male' and 'female' then with likelihood $2pq$, an edge is cross-gender.

Homophily test: if the fraction of cross-gender edges is significantly less than $2pq$, then there is evidence for homophily.

$$(\text{Significant} = \text{Statistically Significant}) \tag{9.2}$$

Homophily test example:

Consider Figure 9.8; a friendship network of a (small) hypothetical classroom in which the three shaded nodes are girls and the six unshaded nodes are boys. Five of the 18 edges in the graph are cross-gender.

Since $p = 6/9 = 2/3$ and $q = 3/9 = 1/3$

we should be comparing the fraction of cross-gender edges to $2pq = \frac{4}{9} = 8/18$. (For no homophily, there should be zero cross-gender edges). But actually only five edges are cross-gender edges,

Thus, homophily exists as $5/18 < 8/18$ $(.278 < 0.44)$ (because the fraction of cross-gender edges is significantly less than $2pq$).

9.8 TRANSITIVITY

Transitivity in social network analysis is a property of ties: if there is a tie between A and B and one between B and C, then in a transitive network, A and C will also be connected. Strong ties are more often transitive than weak ties; transitivity is therefore evidence for the existence of strong ties (but not a necessary or sufficient condition). Transitivity and homophily together lead to the formation of *cliques* (fully connected clusters).

A strong transitivity is one in which there are connections AB, BC and AC, and the connection AC is stronger than the minimum value of a strong tie. A weak transitivity is one in which there are connections AB, BC and AC, and AC; the value of

AC is less than the threshold for a strong tie, but greater than the threshold minimum value of weak tie. Two other methods are also available.

- A Euclidean transitivity is defined as a case where AB, BC *and* AC are present, and AC has a value less than the sum of $AB + BC$.
- A stochastic transitivity is defined as the case where AB, BC *and* AC are present, and AC is less than the produce $AB * BC$.

A node's *clustering coefficient* is the number of closed triplets in the node's neighbourhood over the total number of triplets in the neighbourhood. It is also known as *transitivity*.

Or we can say, transitivity occurs if *triangles* exist in the graph. This can be used to measure the global clustering coefficient as:

$$C = \frac{3 \times \text{Number of triangles in the graph}}{\text{Number of connected triples in the graph}} \tag{9.3}$$

Where a *connected triple* is a triple of nodes consisting of a central node connected to the other two; the flanking nodes are unordered. The equation counts the fraction of connected triples that actually form triangles; the factor of three is due to the fact that each triangle corresponds to three triples. For example, node 1 to the right has a value of 1 because it is only connected to 2 and 3, and these nodes are also connected to one another (i.e., the only triplet in the neighbourhood of 1 is closed). We say that nodes 1, 2 and 3 form a *clique*.

Clustering algorithms identify clusters or 'communities' within networks based on network structure and specific clustering criteria.

$$\text{Network clustering coefficient} = \frac{3 \text{ nodes in each triangle} \times 2 \text{ triangles}}{\text{total triples}}$$

$$= \frac{3 \times 2}{16} = 0.375$$

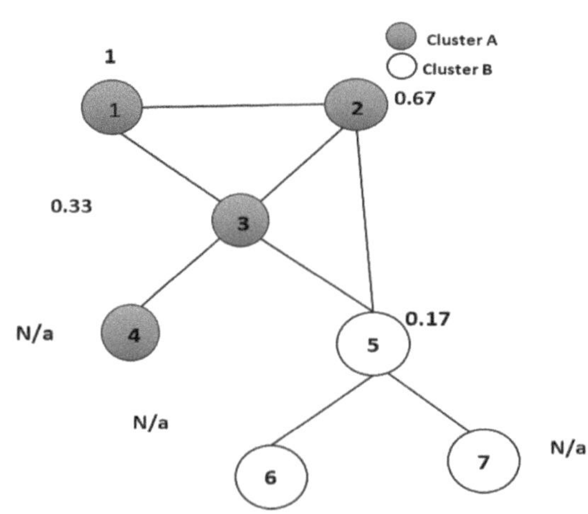

FIGURE 9.9 Example of clustering coefficient

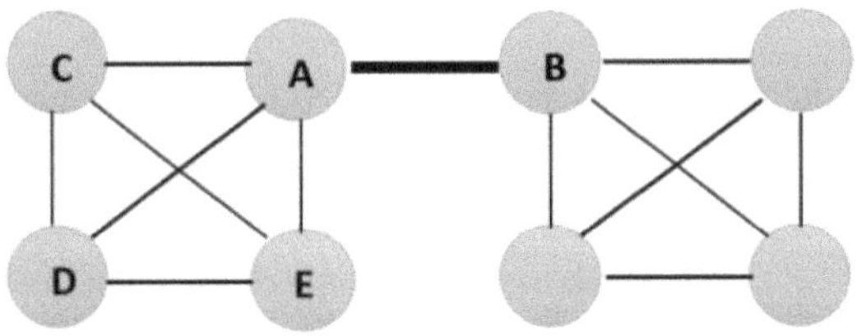

FIGURE 9.10 Bridges example

9.9 BRIDGES

Bridges are nodes and edges that connect across groups. A bridge is a type of social tie that connects two different groups in a social network, facilitates inter-group communication, increases social cohesion and helps spur innovation. They are usually weak ties, but not every weak tie is a bridge. In other words, an edge joining two nodes A and B in a graph is a bridge if deleting the edge would cause A and B to lie in two different components. This edge is literally the only route between its endpoints, the nodes A and B. Common bridge: C A B D E.

In Figure 9.11, the A–B edge isn't the only path that connects its two endpoints; although they may not realize it, A and B are also connected by a longer path through F, G and H. This kind of structure is arguably much more common than a bridge in real social networks, and we use the following definition to capture it.

Local bridge: An edge joining two nodes A and B in a graph is a local bridge if its endpoints A and B have no friends in common – in other words, if deleting the edge would increase the distance between A and B to a value strictly more than two. We say that the span of a local bridge is the distance its endpoints would be from each other if the edge were deleted [1]. Thus, in Figure 9.11, the A–B edge is a local bridge with span four; we can also check that no other edge in this graph is a local bridge, since for every other edge in the graph, the endpoints would still be at distance two if the edge were deleted. Notice that the definition of a local bridge already makes an implicit connection with triadic closure, in that the two notions

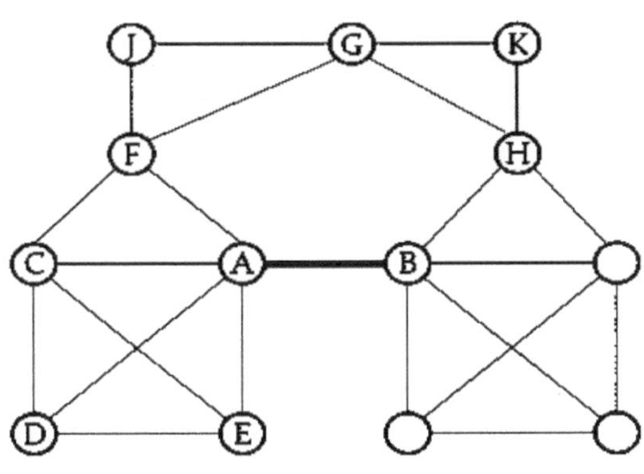

FIGURE 9.11 Example of a bridge

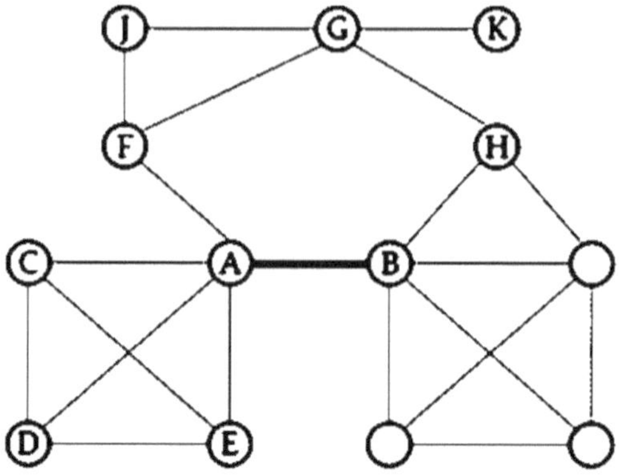

FIGURE 9.12 Example of a local bridge

form conceptual opposites: an edge is a local bridge precisely when it does not form a side of any triangle in the graph [1].

Thus, the A–B edge is a local bridge of span four, since the removal of this edge would increase the distance between A and B to four.

An edge from A to B is called a local bridge if A and B have no friends in common: J G K.

Local bridges are the edges of neighbourhood overlap 0 and hence we can think of edges with very small neighbourhood overlap as being almost local bridges. Intuitively, edges with very small neighbourhood overlap consist of nodes that travel in social circles having almost no one in common.

The neighbourhood overlap of an edge (x, y):

$$\mathrm{NO}(x, y) = \frac{|\text{common neighbours of } x \text{ and } y|}{|\text{neighbours of at least one of } x \text{ or } y|}$$

$$= \frac{|(\mathrm{N}(x) \cap \mathrm{N}(y)|}{|(\mathrm{N}(x) - \{y\}) \cup (\mathrm{N}(y) - \{x\})|}$$

For Figure 9.12, NO (A, B) = 0, and NO (A, F) = 1/6.

9.10 GROUP-EXTERNAL AND GROUP-INTERNAL TIES

Actors may be embedded in macro-structures, as well as in dyads, triads and neighbourhoods. Some macro-structures are social agents (like voluntary and formal organizations) and some macro-structures are categorical units (like gender and ethnic groups). To understand the 'texture' of the 'social fabric', we might want to index the extent to which these macro-structures 'cluster' the interaction patterns of individuals who fall within them.

Krackhardt and Stern developed a very simple and useful measure (E-I Index) of the group embedding based on comparing the numbers of ties within groups and between groups [4].

E-I Index:

The E-I (external-internal) Index takes the number of ties of group members to outsiders, subtracts the number of ties to other group members and divides by the total number of ties. The resulting index ranges from –1 (all ties are internal to the group) to +1 (all ties are external to the group). Since this measure is concerned with any connection between members, the directions of ties are ignored (i.e., either an out-tie or an in-tie constitutes a tie between two actors). Thus, assuming two groups based on some attribute, one defined as internal and the other as external:

$$E-I \text{ Index} = \frac{EL - IL}{EL + IL} \tag{9.4}$$

where,

EL = the number of external friendship links
IL = the number of internal friendship links

Interpretation:

- Score of +1.0 = All links external to subunit
- Score of 0 = Links are divided equally
- Score of –1.0 = All links are internal to subunit

The E-I Index can be applied at three levels: the entire population, each group and each individual.

That is, the network as a whole (all the groups) can be characterized in terms of the bounded-ness and closure of its subpopulations. We can also examine variation across the groups in their degree of closure; and, each individual can be seen as more or less embedded in their group.

Note: the relative size of subpopulations (e.g., 10 vs 1,000) has dramatic consequences for the degree of internal and external contacts, even when individuals may choose contacts at random.

E–I Index example:

$$E-I \text{ Index} = \frac{EL - IL}{EL + IL}$$

For the left portion: $(1-4)/7 = -3/7$; for the right portion: $(1-2)/7 = -1/7$.

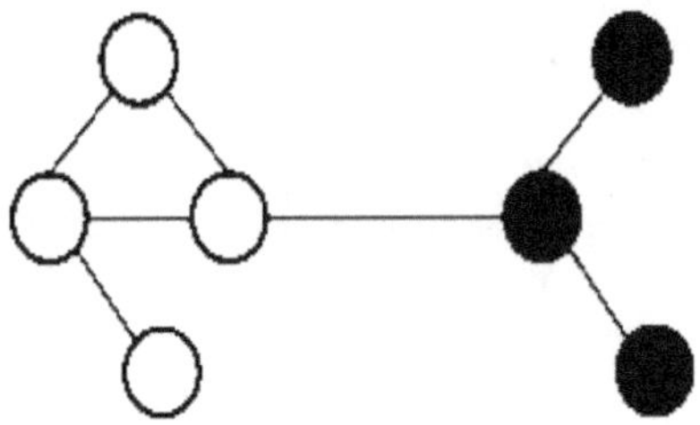

FIGURE 9.13 Example of the E–I Index

9.11 KRACKHARDT'S GRAPH THEORETICAL DIMENSIONS OF HIERARCHY

Embedding of actors in dyads, triads, neighbourhoods, clusters and groups are all ways in which the social structure of a population may display 'texture'. All of these forms of embedding structures speak to the issue of the 'horizontal differentiation' of the population – separate, but not necessarily ranked or unequal groupings.

A very common form of embedding of actors in structures, though, does involve unequal rankings. Hierarchies, in which individuals or subpopulations are not only differentiated, but also ranked, are extremely common in social life. The degree of hierarchy in a population speaks to the issue of 'vertical differentiation'. While we all have an intuitive sense of what it means for a structure to be a hierarchy. Most would agree that structures can be 'more or less' hierarchical. It is necessary to be quite precise about the meaning of the term if we are going to build indexes to measure the degree of hierarchy.

Krackhardt provided an elegant definition of the meaning of hierarchy, and developed measures of each of the four component dimensions of the concept that he identified. Krackhardt defines a pure, 'ideal typical' hierarchy as an 'out-tree' graph. An out-tree graph is a directed graph in which all points are connected, and all but one node (the 'boss') has an indegree of one. This means that all actors in the graph (except the ultimate 'boss') have a single superior node. The simplest 'hierarchy' is a directed line graph A to B to C to D. More complex hierarchies may have wider, and varying 'spans of control' (outdegrees of points) [5].

This very simple definition of the pure type of hierarchy can be deconstructed into four individually necessary and jointly sufficient conditions. Krackhardt develops index numbers to assess the extent to which each of the four dimensions deviates from the pure ideal type of an out-tree, and hence develops four measures of the extent to which a given structure resembles the ideal typical hierarchy.

1. **Connectedness:** To be a pure out-tree, a graph must be connected into a single component – all actors are embedded in the same structure. We can measure the extent to which this is not true by looking at the ratio of the number of pairs in the directed graph that are reachable relative to the number of ordered pairs. That is, what proportion of actors cannot be reached by other actors? Where a graph has multiple components – multiple unconnected subpopulations – the proportion not reachable can be high. If all the actors are connected in the same component, if there is a 'unitary' structure, the graph is more hierarchical.
2. **Hierarchy:** To be a pure out-tree, there can be no reciprocated ties. Reciprocal relations between two actors imply equal status, and this denies pure hierarchy. We can assess the degree of deviation from pure hierarchy by counting the number of pairs that have reciprocated ties relative to the number of pairs where there is any tie; that is, what proportion of all tied pairs have reciprocated ties.
3. **Efficiency:** To be a pure out-tree each node must have an in-degree of one. That is, each actor (except the ultimate boss) has a single boss. This aspect of

the idea type is termed 'efficiency' because structures with multiple bosses have unnecessary redundant communication of orders from superiors to subordinates. The amount of deviation from this aspect of the pure out-tree can be measured by counting the difference between the actual number of links (minus 1, since the ultimate boss has no boss) and the maximum possible number of links. The bigger the difference, the greater the inefficiency. This dimension then measures the extent to which actors have a 'single boss'.

4. **Least upper bound:** To be a pure out-tree, each pair of actors (except pairs formed between the ultimate boss and others) must have an actor that directs ties to both – that is, command must be unified. The deviation of a graph from this condition can be measured by counting the numbers of pairs of actors that do not have a common boss relative to the number of pairs that could (which depends on the number of actors and the span of control of the ultimate boss).

9.12 POSITIONS AND ROLES

The social roles or positions are defined by regularities in the patterns of relations among actors, not attributes of the actors themselves. The 'attributes of individuals' can be thought of as shorthand labels for patterns of relations [6].

Positions: actors that show a similar structure of relationships and are thus similarly embedded into the network.

Roles: the pattern of relationships of members of same or different positions.

Many of the category systems used by sociologists are based on 'attributes' of individual actors that are common across actors.

Approaches to Network Positions and Social Roles

Because 'positions' or 'roles' or 'social categories' are defined by 'relations' among actors, we can identify and empirically define social positions using network data. In an intuitive way, we would say that two actors have the same 'position' or 'role' to the extent that their pattern of relationships with other actors is the same. But there are a couple of things about this intuitive definition that are troublesome.

First, what relations do we take into account, among whom, in seeking to identify which actors are similar and which are not? The relations that someone has with the university (as 'Professor') are similar in some ways to the relations that his students have with the university: we are both governed by many of the same rules, practices, and procedures. The relations someone has with the university are very different from those of his students in some ways (e.g., the university pays me, students pay the university). Which relations should count and which ones not, in trying to describe the roles of 'professor' and 'student?' Indeed, why is someone examining relations among his students, himself and the university, instead of including, say, members of the state legislature?

There is no simple answer about what the 'right relations' are to examine; and, there is no simple answer about who the relevant set of 'actors' are. It all depends upon the purposes of our investigation, the theoretical perspective we are using and the populations to which we would like to be able to generalize our findings. Social network data analytic methods are of little use in answering these conceptual questions.

The second problem with our intuitive definition of a 'role' or 'position' is this: assuming that someone has a set of actors and a set of relations that make sense for studying a particular question, what does it mean that actors who share the same position are similar in their pattern of relationships or ties? The idea of 'similarity' has to be rather precisely defined. Again, there is no single and clear 'right' answer for all purposes of investigation. But there are rigorous ways of thinking about what it means to be 'similar' and there are rigorous ways of actually examining data to define social roles and social positions empirically. These are the issues where there are some ways in which widely used methods can provide guidance. The 'positions' or 'roles' or 'social categories' are defined by 'relations' among actors. Two actors have the same 'position' or 'role' to the extent that their pattern of relationships with other actors is the same.

Similarity: the idea of 'similarity' has to be rather precisely defined as nodes are similar if they fall in the same 'equivalence class'.

Equivalence: there are three particular definitions of equivalence:

1. **Structural equivalence:** two nodes are said to be exactly structurally equivalent if they have the same relationships to all other nodes.
2. **Automorphic equivalence (rarely used):** the idea of automorphic equivalence is that sets of actors can be equivalent by being embedded in local structures that have the same patterns of ties – 'parallel' structures.
3. **Regular equivalence:** two nodes are said to be regularly equivalent if they have the same profile of ties with members of other sets of actors that are also regularly equivalent.

Example of Equivalence Classes

Suppose that the graph in Figure 9.14 is a franchise group of hamburger restaurants.

- Seven structural equivalence classes: {A}, {B}, {C}, {D}, {E, F}, {G} and {H, I}.
  Explanation: there is no actor who has exactly the same set of ties as actor A (ties to B, C and D), so actor A is in a class by itself. The same is true for actors B, C and D. Each of these actors has a unique set ties to others, so they form three classes, each with one member. E and F, however, fall in the same structural equivalence class. Each has a single tie; and that tie

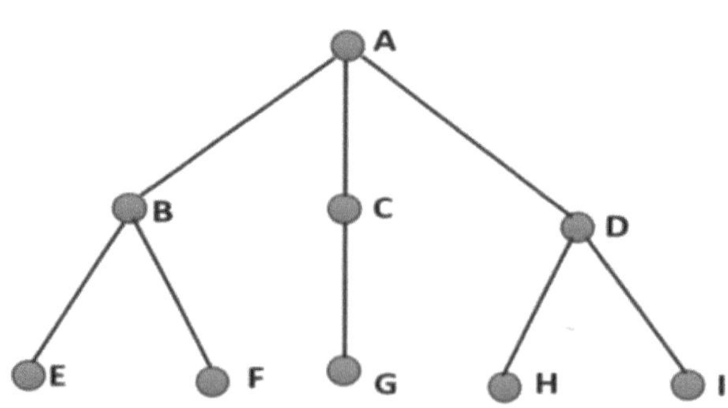

FIGURE 9.14 Example graph for equivalence

is to actor B. Since E and F have exactly the same pattern of ties with all other actors, they are structurally equivalent. Actor G, again, is in a class by itself. Its profile of ties with the other nodes in the diagram is unique. Finally, actors H and I fall in the same structural equivalence class. That is, they have exactly the same pattern of ties to all other actors.

- Five automorphic equivalence classes: {A}, {B, D}, {C}, {E, F, H, I} and {G}
Explanation: actor A is the central headquarters, actors B, C and D are the managers of three different stores. Actors E and F are workers at one store; G is the lone worker at a second store; H and I are workers at the third store. Even though actor B and actor D are not structurally equivalent (they do have the same boss, but not the same workers), they do seem to be 'equivalent' in a different sense. Both manager B and D report to a boss (in this case, the same boss), and each has exactly two workers. These are different people, but the two managers seem somehow equivalent. If we swapped them, and also swapped the four workers, all of the distances among all the actors in the graph would be exactly identical. In fact, actors B and D form an 'automorphic' equivalence class.
- Three regular equivalence classes: {A}, {B, C, D} and {E, F, G, H, I}.
Explanation: the first is actor A; the second is composed of the three actors B, C and D; the third is composed of the remaining five actors E, F, G, H and I. The easiest class to see is the five actors across the bottom of the diagram (E, F, G, H and I). These actors are regularly equivalent to one another because (1) they have no tie with any actor in the first class (that is, with actor A) and (2) each has a tie with an actor in the second class (either B or C or D). Each of the five actors, then, has an identical pattern of ties with actors in the other classes. Actors B, C and D form a class because (1) they each have a tie with a member of the first class (that is, with actor A) and (2) they each have a tie with a member of the third class. B and D actually have ties with two members of the third class, whereas actor C has a tie to only one member of the third class; this doesn't matter, as there is a tie to some member of the third class. Actor A is in a class by itself, defined by (1) a tie to at least one member of class two and (2) no tie to any member of class three.

As with structural and automorphic equivalence, exact regular equivalence may be rare in a large population with many equivalence classes [7]. Approximate regular equivalence can be very meaningful though, because it gets at the notion of which actors fall in which social roles, and how social roles (not role occupants) relate to one another. The definitions of the forms of equivalence discussed here are quite precise

The notions of equivalence provide quite rigorous ways of defining and thinking about core analytical tools in sociology – individual's positions in groups, types of structures, and social roles. This is a huge advance over the sometimes quite imprecise and contradictory verbal treatments found in much of our literature. But real-world social networks are often quite messy, may not be fully realized (that is, not in equilibrium) and/or may be badly measured. The search for equivalence in real data can be a somewhat complicated matter with a number of vexing choices to be made.

9.13 PROBLEMS FOR SELF-ASSESSMENT

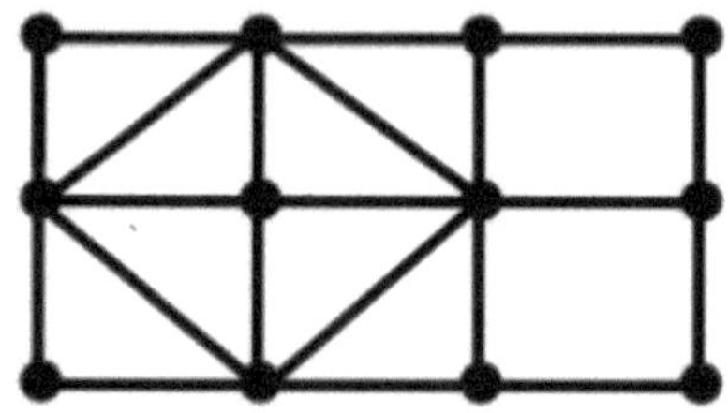

Q1. What is the highest and lowest embeddedness value among the edges in the network shown in the graph below?

Q2. I have four friends. If my clustering coefficient is 0.33, what is the number of friendships between my friends?

Q3. Does this network below exhibit homophily? If yes, to what extent does it exhibit? If no, why?

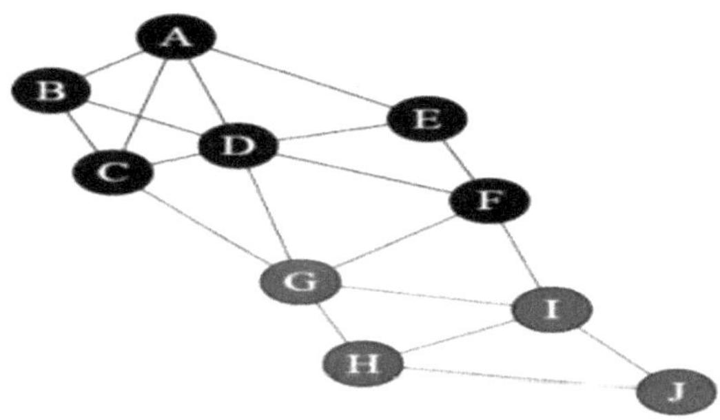

Q4. Consider the graph below and find the all the equivalence classes of each category (structural, automorphic and regular equivalence).

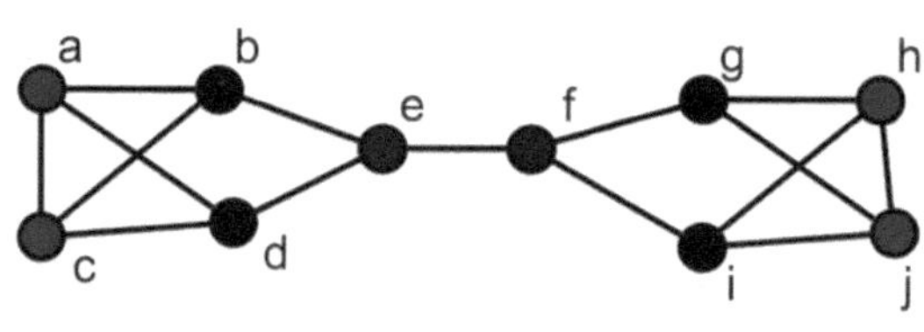

REFERENCES

1. Easley, D., & Kleinberg, J. (2010). Strong and weak ties. *Networks, Crowds, and Markets: Reasoning about a Highly Connected World*. Cambridge University Press. https://www.cs.cornell.edu/home/kleinber/networks-book/networks-book-ch03.pdf
2. Punyachonkool, C., & Sakworawich, A. (2016, Sep. 1). *Subscriber churn prediction model using social network analysis in telecommunication*. https://www.slideshare.net/BAINIDA/subscriber-churn-prediction-model-using-social-network-analysis-in-telecommunication-industry
3. Hanneman, R., & Riddle, M. (2005). Embedding: Krackhardt's graph theoretical dimensions of hierarchy. *Introduction to Social Network Methods*. http://faculty.ucr.edu/~hanneman/nettext/C8_Embedding.html

4. Krackhardt, D., & Stern, R. N. (1988). Informal networks and organizational crises: An experimental simulation. *Social Psychology Quarterly*, 51, 123–140.
5. Krackhardt, D., & Brass, D. J. (1994). Intraorganizational networks: The micro side. In: Wasserman, S. & Galaskiewicz, J. eds, *Advances in Social Network Analysis: Research in the Social and Behavioral Sciences*, Sage.
6. Hanneman, R., & Riddle, M. (2005). Network positions and social roles: The idea of equivalence. *Introduction to Social Network Methods.* https://faculty.ucr.edu/~hanneman/nettext/C12_Equivalence.html
7. Hanneman, R., & Riddle, M. (2005). Defining equivalence or similarity. *Introduction to Social Network Methods.* http://faculty.ucr.edu/~hanneman/nettext/C12_Equivalence.html#define

10 Information Diffusion

10.1 OVERVIEW OF INFORMATION DIFFUSION

Today, both sharing and spreading any information online have become an integral part of modern life, for example people share day-to-day family activities, forward news articles to their relatives and friends, transfer YouTube videos to their near ones and send movie or restaurant feedback/reviews to their neighbours [1]. All that people want is to make their content popular/diffused/viral over online platforms so that they can gain attention.

Thus, diffusion of information over social media has become a most curious area. Information diffusion/virality is defined as a process by which the posted content/information (knowledge) spreads and reaches individuals through interactions and connections. The terms content virality and information diffusion can be used interchangeably in the context of social networks.

Information diffusion or virality over online social networks raises the following questions [2]:

- What are the main topics and the part of content/information that become popular and diffuse/spread the most?
- How is information diffused, why is information diffusing and through which paths is information diffusing?
- Will the information be diffused in the future?
- Which people in the network/connections are the most important people who play a vital role in the spreading process?

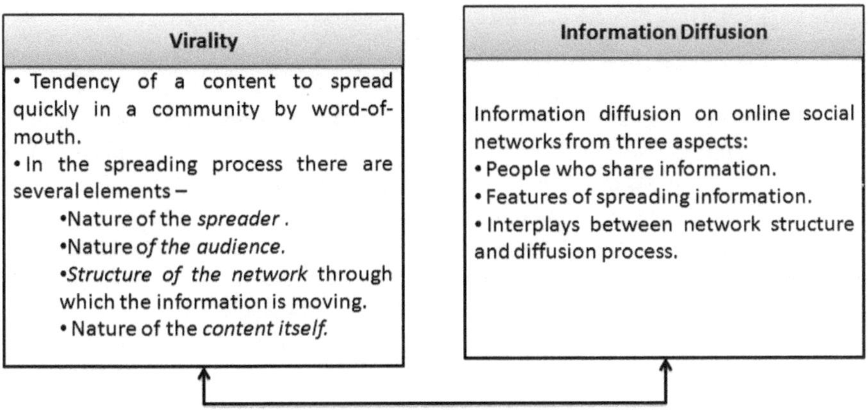

FIGURE 10.1 Virality versus information diffusion

DOI: 10.1201/9781003088066-10

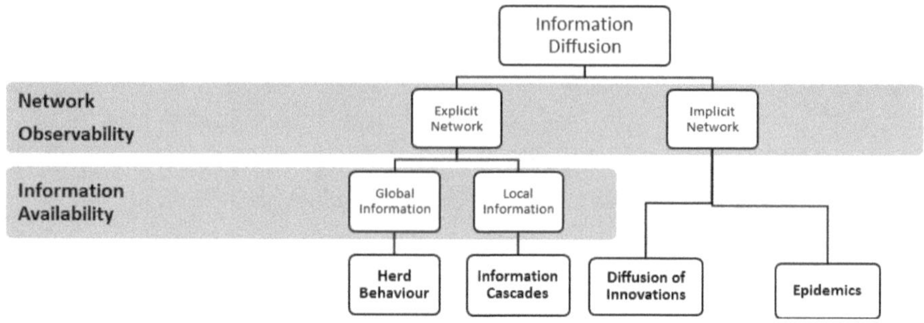

FIGURE 10.2 Types of information diffusion [3]

Three main elements are involved in the diffusion process: sender, receiver and the medium. **Sender** is the one who starts the diffusion process. **Receiver** is the one who gets/sees the diffused information/content. And the **medium** is the platform through which the spread of information occurs. For example, the spread of a rumour can be through personal communication between individuals. Types of information diffusion on social media platforms can be defined on the basis of network observability and information availability during diffusion. Therefore, information diffusion on social media can have four general types:

1. Herd behaviour
2. Information cascades
3. Diffusion of innovations
4. Epidemics

During information diffusion on social media, two forms of the network can be observed: explicit networks and implicit networks.

10.2 EXPLICIT NETWORKS

In explicit networks, connections among social media users are determined explicitly by the users themselves to whom they should be connected to [4]. Users issue and accept friendship requests on Facebook. On Twitter, likewise, where netizens decide that they want to follow the tweets of explicit/specific users.

In all cases, the topology of the subsequent network reflects the selections of users and frequently comprises of connections that as of now exist between genuine individuals. To reinforce and exploit the existing connections, explicit networks are very useful. These networks also provide little support for discovering new content over social media platforms [4].

Information Availability: The type of diffusion of information on social platforms depends on the availability of the information to individuals. These individuals are situated in a network where they interact with each other. Although the network, on which the information diffusion takes place, is at times unavailable or imperceptible. The flow of information over the network depends on the individual's decision. For example, when a rumour is spreading, individuals decide if they are interested in spreading it to their neighbours. This decision can be taken independently or dependently relying on the information they receive. When the decision of information spread is dependent on the others, it is very important to see the intensity of the dependency. Dependency can be local, where spread of information depends on the immediate neighbours, or it can be global, where the spread of information depends on the whole network [3].

Global Information: On the basis of global information, diffusion can be seen in the form of herd behaviour.

10.2.1 Herd Behaviour

When the spread of information depends on the whole network (i.e., on global information), herd behaviour takes place [3].

Herd Behaviour (Popular Restaurant Experiment): Assume that you are on a trip in a metropolitan area that you are less familiar with. Planning for dinner, you find restaurant A with excellent reviews online and decide to go there. When arriving at A, you see A is almost empty and restaurant B, which is next door and serves the same cuisine, almost full. Deciding to go to B, based on the belief that other diners have also had the chance of going to A, is an example of herd behaviour. Herd behaviour describes when a group of individuals performs actions that are highly correlated without any plans.

Main components of herd behaviour:

- A method to transfer behaviour among individuals or to observe their behaviour.
- A connection between individuals.

Examples of Herd Behaviour: Flocks, herds of animals, humans during sporting events, demonstrations and religious gatherings. In herd behaviour, individuals make decisions by observing all other individuals' decisions.

In general, herd behaviour's network is close to a complete graph where nodes can observe at least most other nodes and they can observe public information. For example:

- They can see the crowd.
- There needs to be a decision made.
- In our example, it is going to a restaurant.
- Decisions need to be in sequential order.
- Decisions are not mindless and people have private information that helps them decide.
- No message passing is possible. Individuals don't know the private information of others, but can infer what others know from what is observed from their behaviour.

Herding Intervention: In herding, the society only has access to public information. Herding may be intervened by releasing private information which was not accessible before.

Local Information: On the basis of the local information, diffusion takes place in the form of cascades.

10.2.2 Information Cascades

When the spread of information depends on the immediate neighbours (i.e., on local information), information cascading takes place [2]. In social media, individuals commonly repost content posted by others in the network. This content is often received via immediate neighbours (friends). An information cascade occurs as information propagates through friends. An information cascade is defined as a piece of information or decision being cascaded among a set of individuals, where individuals are connected by a network and individuals are only observing decisions of their immediate neighbours (friends).

Therefore, cascade users have less information available to them compared to herding users, where almost all information about decisions are available. In cascading, local information is available to the users, but in herding, the information about the population is available.

Underlying Assumptions for Cascade Models

The network is represented using a directed graph. Nodes are actors and edges depict the communication channels between them. A node can only influence nodes that it is connected to:

- Decisions are binary – nodes can be either active or inactive. An active node means that the node decided to adopt the behaviour, innovation or decision.
- A node, once activated, can activate its neighbouring nodes.
- Activation is a progressive process, where nodes change from inactive to active, but not vice versa.

Independent Cascade Model

The Independent Cascade Model (ICM) is a sender-centric model of a cascade. In this model, each node has one chance to activate its neighbours. Considering nodes that are active as senders and nodes that are being activated as receivers, the *linear threshold model* concentrates on the receiver. The independent cascade model concentrates on the sender. In an ICM, the node that is activated at time t, has one chance, at time step $t + 1$, to activate its neighbours.

- Let v be an active node at time t, for any neighbour w of it, there's a probability p_{vw} that node w gets activated at time $t + 1$.
- A node v activated at time t has a single chance of activating its neighbours and that activation can only happen at $t + 1$

ICM:

Requires: Diffusion graph G (V, E), set of initial activated nodes A_0 and activation probabilities $p_{v,w}$

1. **return** Final set of activated nodes A_∞
2. $i = 0$;
3. **while** $A_i \neq \{\}$ *d*
4.
5. $i = i + 1$;
6. $A_i = \{\}$;
7. **for all** $v \in A_{i-1}$ **do**
8. **for all** w neighbor of $v, w \notin \bigcup_{j=0}^{i} A_i$ **do**
9. Rand = generate a random number in $[0,1]$;
10. **if** rand $< p_{v,w}$ then
11. Activate w_i;
12. $A_i = A_i \cup \{w\}$;
13. **end if**
14. **end for**
15. **end while**
16. $A_\infty = \bigcup_{j=0}^{i} A_j$;
17. return A_∞

Maximizing the Spread of Cascades

Maximizing the spread of cascades is the problem of finding a small set of nodes in a social network such that their aggregated spread in the network is maximized. The applications are product marketing and influence.

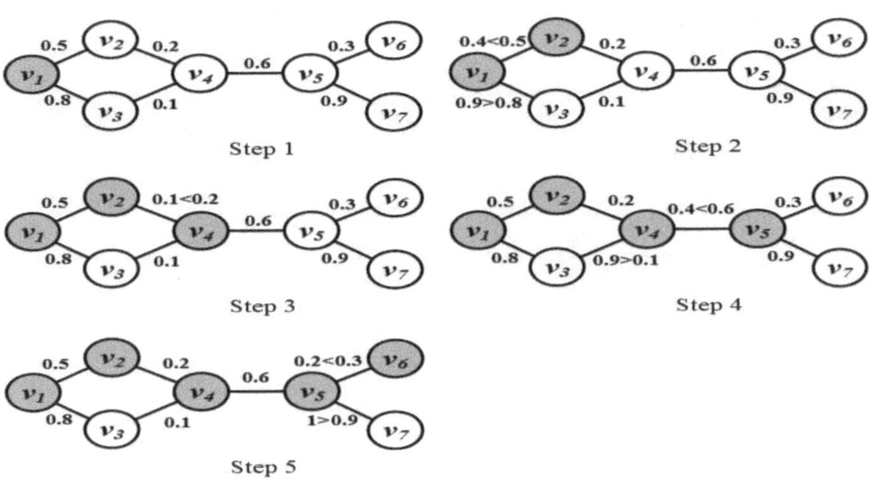

FIGURE 10.3 ICM example [3]

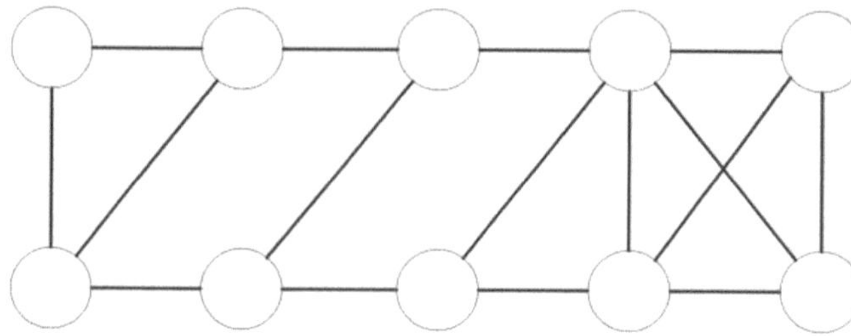

FIGURE 10.4 Example graph for maximizing the spread of cascades

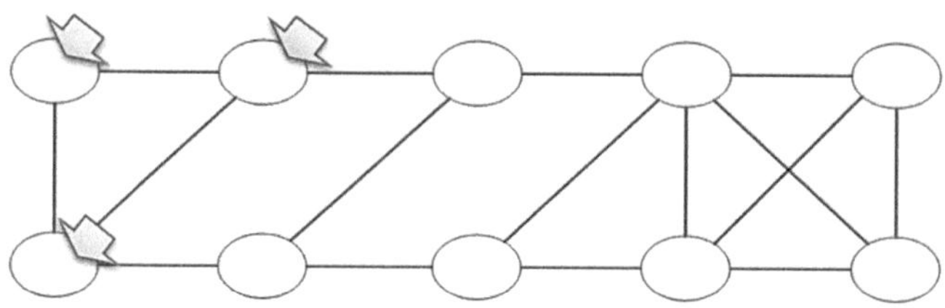

FIGURE 10.5 Information spread after selecting one seed

Problem setting: if it is given – A limited budget B for initial advertising (e.g., give away free samples of product) and estimating spread between individuals. The goal is to trigger a large spread (e.g., further adoptions of a product).

The problem is which set of individuals should be targeted at the very beginning?

Maximizing the spread of cascades example:

We need to pick k nodes such that maximum number of nodes are activated.

Select one seed

Select two seeds

Problem statement: we have to spread the information to node set $S: f(S)$. An expected number of active nodes, if set S is the initial active set. Thus, for given a parameter k (budget), find a k-node set S to maximize $f(S)$. A constrained optimization problem with *f(S)* as the objective function.

$f(S)$ **properties:** non-negative (obviously), monotone, submodular; Let N be a finite set, a set function is submodular if:

$$f : 2^N \mapsto R;\ \forall S \subset T \subset N,\ \forall v \in N \backslash T;\ f(S+v) - f(S) \geq f(T+v) - f(T) \quad (10.1)$$

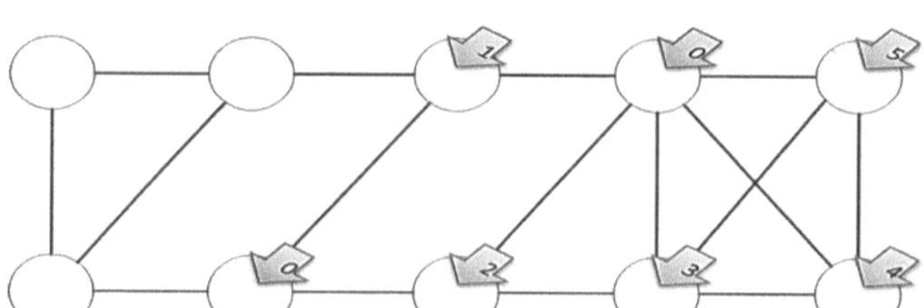

FIGURE 10.6 Information spread after selecting two seeds

Some facts regarding this problem:

Bad news: for a submodular function monotone non-negative *f*, finding a *k*-element set *S* for which *f(S)* is maximized is an NP-hard optimization problem. It is NP-hard to determine the optimum for influence maximization for both independent cascade model and linear threshold model (to be introduced in next chapter).

Good news: we can use a greedy algorithm. Start with an empty set *S*. For *k* iterations, we add node v to S that maximizes $f(S+v)-f(S)$.

Cascade Maximization – A Greedy Approach

Maximizing the cascade is an NP-hard problem but it is proved that the greedy approaches give a solution that is at least 63% of the optimal.

Given a network and a parameter *k*, which *k* nodes should be selected to be in the activation set B in order to maximize the cascade in terms of the total number of active nodes. Let $\sigma(B)$ denote the expected number of nodes that can be activated by B, the optimization problem can be formulated as follows:

$$\max_{B\subseteq V} \sigma(B)\, s.t. |B| \leq k \tag{10.2}$$

ALGORITHM: MAXIMIZING THE SPREAD OF CASCADES

Requires: diffusion graph G (V, E), budget *k*

1. Return seed set *S* (set of initially activated nodes)
2. $i = 0$;
3. $S = \{\}$;
4. While $i \neq k$ do
5. $v = arg \max_{v\in V/S} f(S\cup\{v\})$; or equivalently $arg \max_{v\in V/S} f(S\cup\{v\}) - f(s)$;
6. $S = S\cup\{v\}$;
7. $i = i+1$;
8. end while
9. Return S;

Example: for the following graph, assume that node *I* activates node *j* when $|i-j| \equiv 2(mod3)$. Solve cascade maximization for $k=2$. [3]

To find the first node v, we compute $f(\{v\})$ for all v. We start with node 1. At time 0, node 1 can only activate node 6, because

$$|1-6| \equiv 2(\text{mod}3)$$

At time 1, node 1 can no longer activate others, but node 6 is active and can activate others. Node 6 has outgoing edges to nodes 4 and 5. From 4 and 5, node 6 can only activate 4:

$$|6-4| \equiv 2(\text{mod}3)$$

$$|6-5| \not\equiv 2(\text{mod}3)$$

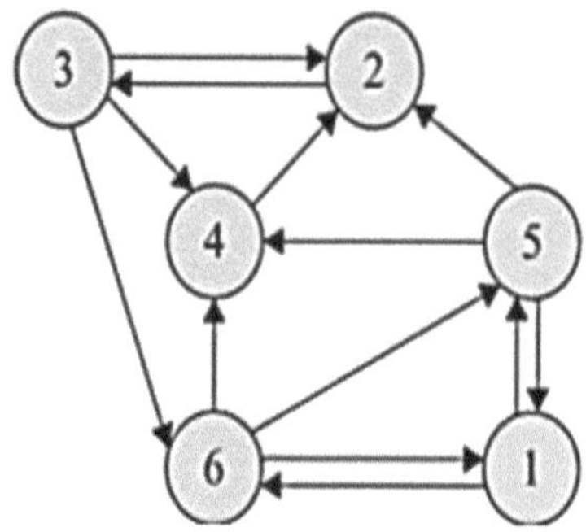

FIGURE 10.7 Example of cascade maximization [3]

At time 2, node 4 is activated. It has a single out-link to node 2 and since $|4-2| \equiv 2(mod3)$, 2 is activated. Node 2 cannot activate other nodes; therefore, $f(\{1\}) = 4$. Similarly, we find that $f(\{2\}) = 1$, $f(\{4\}) = 2$, $f(\{5\}) = 1$, and $f(\{6\}) = 4$. So, 1 or 6 can be chosen for our first node. Let us choose 6.

If 6 is initially activated, nodes 1, 2, 4 and 6 will become activated at the end. Now, from the set $\{1, 2, 3, 4, 5, 6\} \setminus \{1, 2, 4, 6\} = \{3, 5\}$, we need to select one more node. This is because in the setting for this example, $f(\{6, 1)\} = f(\{6, 2)\} = f(\{6, 4)\} = f(\{6)\} = 4$. In general, one needs to compute $f(S \cup \{v\})$ for all $v \in V \setminus S$. We have $f(\{6, 3\}) = f(\{6, 5\}) = 5$, so we can select one node randomly. We choose 3. So, $S = \{6, 3\}$ and $f(S) = 5$.

10.3 IMPLICIT NETWORKS

Through implicit networks, users can find/discover new content and relations/acquaintances [4]. Dynamic communities are formed in implicit networks with the help of collateral activities of netizens, such as browsing websites, clicks on similar interest products or tagging posts/documents, URLs or images/pictures. In dynamic communities, a user may or may not be aware of the other users of their own communities. Whether the users are aware or not of other members, these dynamic communities have the ability to establish new social relations/connections and provide a way to discover new and useful data. With implicit networks, information diffusion can be seen in two types: diffusion of innovations and epidemical models.

10.3.1 Diffusion of Innovations

In this type of information diffusion process, the network is not observable. Only public information is observable. An innovation is 'an idea, practice or object that is perceived as new by an individual or other unit of adoption' (see https://courses.lumenlearning.com/boundless-marketing/chapter/the-spread-of-new-products/). The theory of diffusion of innovations aims to answer why and how these innovations spread. It also describes the reasons behind the diffusion process, individuals involved, as well as the rate at which ideas spread. In other words, diffusion of

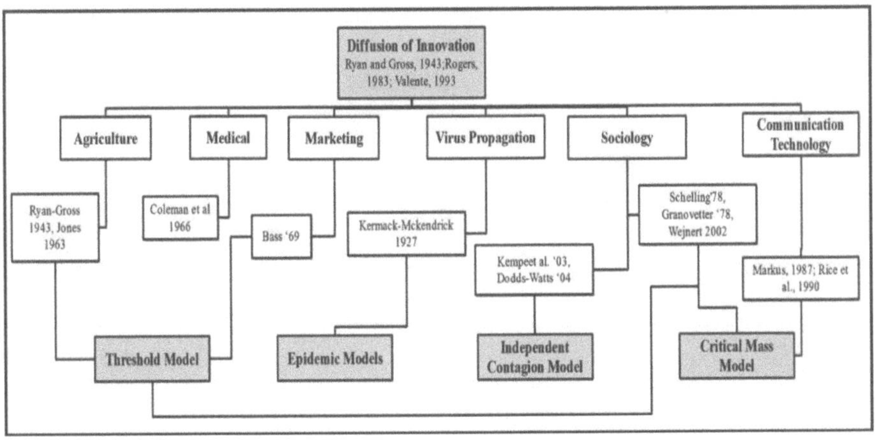

FIGURE 10.8 Various fields and models for diffusion of innovation

innovation provides an outline of how an idea (be it a product or service) diffuses in the crowd. This framework assumes rate of adoption as the sole criteria for governing the process and lays very less impact on individual interactions.

Diffusion of Innovations Models

It is a well-known fact that diffusion is a social process and social strength of peers led to adoption of some information. Predictive models aim to model diffusion and model adoption depends on the social strength-based adoption probability. There are various fields such as agriculture, medical, marketing, virus propagation, sociology, communication technology among others on which these models had been applied by various researchers to examine the diffusion of information over the larger networks. The first model was introduced by Gabriel Tarde in the early 20th century.

The Iowa Study of Hybrid Corn Seed: Ryan and Gross studied the adoption of hybrid seed corn by farmers in Iowa. Farmers received information through two main channels: mass communications from companies selling the seeds (information) and interpersonal communications with other farmers (influence). They argued that adoption depended on a combination of both. They also observed that the adoption rate

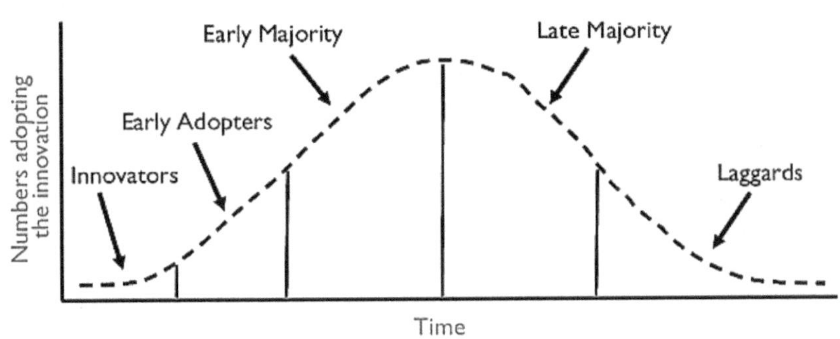

FIGURE 10.9 Adopters categories [5]

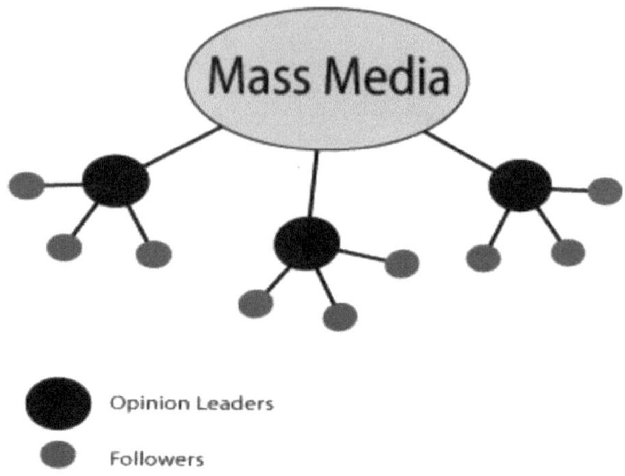

FIGURE 10.10 Two-step flow model of diffusion [6]

follows an S-shaped curve and that there are five different types of adopters based on the order that they adopt the innovations, namely: Innovators (top 2.5%), Early Adopters (13.5%), Early Majority (34%), Late Majority (34%) and Laggards (16%).

Despite the fact that the use of new seed could lead to an increase in quality and production, the adoption by Iowa farmers was slow. The hybrid corn was highly resistant to diseases and other catastrophes such as droughts. However, farmers did not adopt it due to its high price and its inability to reproduce (e.g., new seeds have to be purchased from the seed provider).

Two-step (multiple-step) flow model of diffusion: according to the two-step flow model, most information comes from mass media, which is then directed toward influential figures called *opinion leaders*. These leaders then convey the information (or form opinions) and act as hubs for other members of the society.

Modelling diffusion of innovations: this diffusion of innovation model describes the rate at which the number of adopters changes in terms of time:

$$\frac{dA(t)}{dt} = i(t)\left[P - A(t)\right] \tag{10.3}$$

where,

$A(t)$ is the total population that adopted the innovation

$i(t)$ denotes the coefficient of diffusion corresponding to the innovativeness of the product being adopted

P is the total number of potential adopters (till time t)

The rate depends on how innovative the product is and the rate affects the potential adopters that have not yet adopted the product.

Mathematical model of information diffusion:

$$\frac{dA(t)}{dt} = i(t)\left[P - A(t)\right] \rightarrow A(t)\int_{t_0}^{t} a(t)dt, \tag{10.4}$$

Defining the diffusion coefficient by defining $i(t)$ as a function of number of adopters $A(t)$, (A_0: the number of adopters at time t_0 and $a(t)$: adopters at time t)

$$i(t)=\alpha+\alpha_0 A_0+\ldots+\alpha_t A(t)=\alpha+\sum_{i=t_0}^{t}\alpha_i A(i) \tag{10.5}$$

Diffusion models: there are three models of diffusion:

- External – Influence Model: $i(t)=\alpha$ (10.6)
- Internal – Influence Model: $i(t)=\beta A(t)$ (10.7)
- Mixed – Influence Model: $i(t)=\alpha+\beta A(t)$ (10.8)

where α is the innovativeness factor of the product and β is the imitation factor.

External-influence model: the adoption rate is a function that depends on external entities, $i(t)=\alpha$

$$\frac{dA(t)}{dt}=\alpha[P-A(t)]\rightarrow A(t)=P\left(1-e^{-\alpha t}\right) \tag{10.9}$$

The number of adopters increases exponentially and then saturates near P.

Internal-influence model: the adoption rate is a function that depends only on the number of already activated individuals, $i(t) = \beta A(t)$

$$\frac{dA(t)}{dt}=\beta A(t)[P-A(t)]\rightarrow A(t)=\frac{P}{1+\frac{P-A_0}{A_0}e^{\beta P(t-t_0)}} \tag{10.10}$$

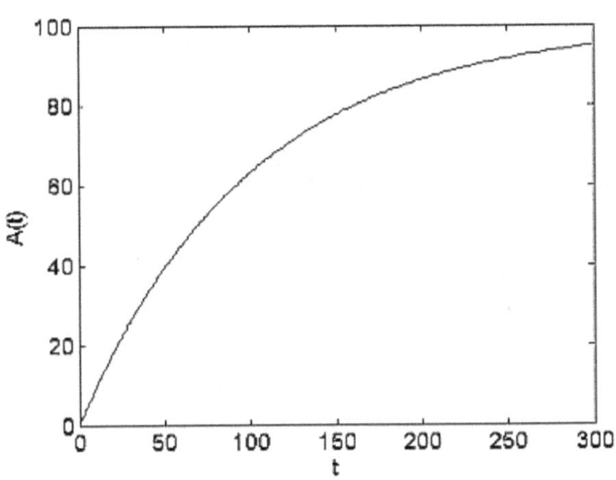

FIGURE 10.11 External-influence model

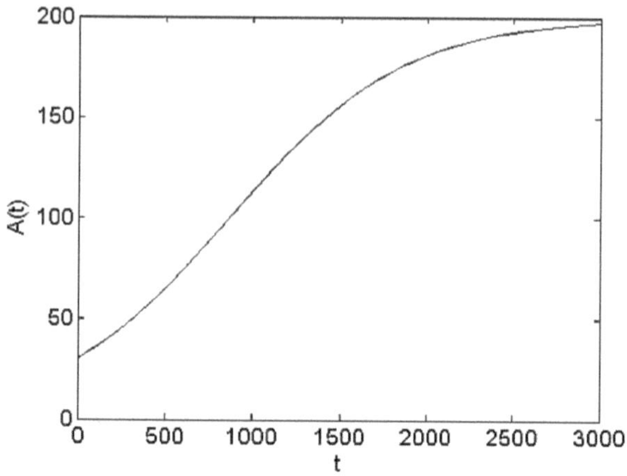

FIGURE 10.12 Internal-influence model

Mixed-influence model: the adoption rate is a function that depends on both the number of already activated individuals and external forces, $i(t) = \alpha + \beta A(t)$

$$\frac{dA(t)}{dt} = \alpha + \beta A(t)A(t)\left[P - A(t)\right] \tag{10.11}$$

Diffusion of innovation – intervention:

- Limiting the distribution of the product or the audience that can adopt the product.
- Reducing interest in the product being sold. For instance, the company can inform adopters of the faulty status of the product.
- Reducing interactions within the population. Reduced interactions result in less imitations on product adoptions and a general decrease in the trend of adoptions.

Case Study to Understand Information Diffusion through YouTube View Counts

Aggrawal et al. (2018) presented an analogy from marketing science and proposed a framework that characterized the literary theory in terms of the view count (The View Count Model) [7]. Their viewership computational model considered the association of two types of viewers in contributing to the overall view count. In their approach, they assumed viewership to be initiated by certain number of viewers after the launch of the video in the internet market. Furthermore, they assumed the rate of viewing any video at a given time to comprise of two components that administrate the viewing process; the first factor constitutes the videos watched through external influence with an impact rate v_1 and the second factor represents the additional number of viewers who watch a video and under the influence of word of mouth (with rate v_2). Their View Count Process can be modelled in the following manner:

$$\frac{dV(t)}{dt} = v_1(N - V(t)) + v_2\frac{V(t)}{N}(N - V(t)) \tag{10.12}$$

Here, v_1 represents the fraction of all viewers who are neoteric. The product v_2/N times $V(T)$ reflects the pressure operating on followers as the number of early viewer's increases. After solving equation (10.12), the closed form solution can be obtained as follows:

$$V(t) = N\left(\frac{1 - e^{-(v_1+v_2)t}}{1 + \frac{v_2}{v_1}e^{-(v_1+v_2)t}}\right) \tag{10.13}$$

Their model so obtained had the ability to act as a forecasting tool that can help the firm in knowing the manner in which a video is able to generate popularity in terms of view count.

10.3.2 Epidemical Models

These models work on a similar platform with a little difference in their framework. Under this ideology, the adoption is replaced by infection and the innovation is dealt in with respect to pathogen. Another differentiation is that, in epidemic models, it is not on individuals to decide whether or not to be infected and it is considered as rather natural process. [8].

Epidemics describes the process by which diseases spread. This process consists of a pathogen (the disease being spread), a population of hosts (humans, animals, plants etc.) and a spreading mechanism (breathing, drinking, sexual activity etc.).

Comparing Epidemics and Cascades

Unlike information cascades and herding and similar to diffusion of innovation models, epidemic models assume an implicit network and unknown connections between individuals. This makes epidemic models more suitable when we are interested in global patterns, such as trends and ratios of people getting infected, and not in who infects whom.

How to Analyze Epidemics?

Contact network: look at how hosts contact each other and devise methods that describe how epidemics happen in networks. A contact network is a graph where nodes represent the hosts and edges represent the interactions between these hosts. For instance, in the case of the HIV/AIDS, edges represent sexual interactions, and in the case of influenza, nodes that are connected represent hosts that breathe the same air.

Fully-mixed: analyze only the rates at which hosts get infected and recover and avoid considering network information.

The models discussed here will assume that no contact network information is available and the process by which hosts get infected is unknown.

Susceptible-Infected (SI) Model

In this model, the *susceptible* individuals get infected and once *infected* they will never get cured.

- **Susceptible:** when an individual is in the susceptible state, they can potentially get infected by the disease.
- **Infected:** an infected individual has the chance of infecting susceptible parties.

Notations: N: size of the crowd, $S(t)$: number of susceptible individuals at time t

- $s(t)=\frac{S(t)}{N}$ (10.14)
- $I(t)$: number of infected individuals at time t; $i(t)=\frac{I(t)}{N}$ (10.15)
- β: Contact probability
- *if* $\beta = 1$ everyone comes to contact with everyone else
- *if* $\beta = 0$ no one meets another individual

$$N = S(t) + I(t) \tag{10.16}$$

At each time stamp, an infected individual will meet βN people on average and will infect βS of them. Since I are infected, βIS will be infected in the next time-step.

SI Model equations:

$$\frac{dS}{dt} = -\beta IS, \frac{dI}{dt} = \beta IS \rightarrow \frac{dI}{dt} = \beta I(N-I) \rightarrow I(t) = \frac{NI_0 e^{\beta t}}{N + I_0\left(e^{\beta t} - 1\right)} \tag{10.17}$$

I_0 is the number of individuals infected at time 0.

Susceptible-Infected-Recovery (SIR) Model

In the SIR model, in addition to the I and S states, a recovery state R is present. In the SIR model, individuals get infected, then some recover. Once hosts recover (or are removed), they can no longer get infected and are not susceptible any longer.

SIR model equations:

$$I + S + R = N \tag{10.18}$$

$$\frac{dS}{dt} = -\beta IS, \tag{10.19}$$

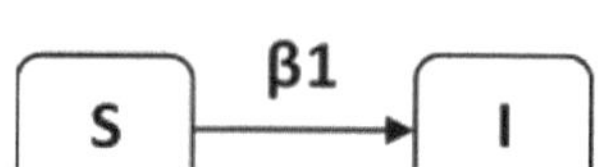

FIGURE 10.13 SI model

FIGURE 10.14 SIR model

$$\frac{dI}{dt} = \beta IS - \gamma I, \tag{10.20}$$

$$\frac{dR}{dt} = \gamma I \tag{10.21}$$

γ defines the recovering probability of an infected individual at a time stamp

$$\frac{dS}{dt} = -\beta IS,$$

$$\frac{dI}{dt} = \beta IS - \gamma I, \rightarrow \frac{dS}{dR} = -\frac{\beta}{\gamma} S \rightarrow log \frac{S_0}{S} = \frac{\beta}{\gamma} R \tag{10.22}$$

$$\frac{dR}{dt} = \gamma I \rightarrow \gamma(N - S - R) \tag{10.23}$$

$$\frac{dR}{dt} = \gamma\left(N - S_0\, e^{-\frac{\beta}{\gamma}R} - R\right) \rightarrow t = \frac{1}{\gamma}\int_0^R \frac{dx}{N - S_0 e^{-\frac{\beta}{\gamma}x} - x} \tag{10.24}$$

There is no closed form solution for this integration and only numerical approximation is possible.

Susceptible-Infected-Susceptible (SIS) Model

The SIS model is the same as the SI model with the addition of infected nodes recovering and becoming susceptible again.

$$\frac{dS}{dt} = \gamma I - \beta IS,\ \frac{dI}{dt} = \beta IS - \gamma I \rightarrow \beta I(N - I) - \gamma I = I(\beta N - \gamma) - \beta I^2 \tag{10.25}$$

- When $\beta N <= \gamma$: the first term will be at most zero or negative hence the whole term becomes negative and therefore, in the limit, the value I(t) will decrease exponentially to zero
- When $\beta N > \gamma$: we will have a logistic growth function like the SI model

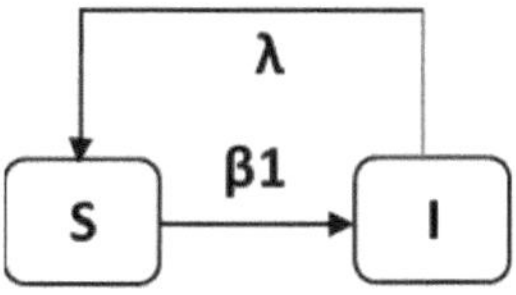

FIGURE 10.15 SIS model

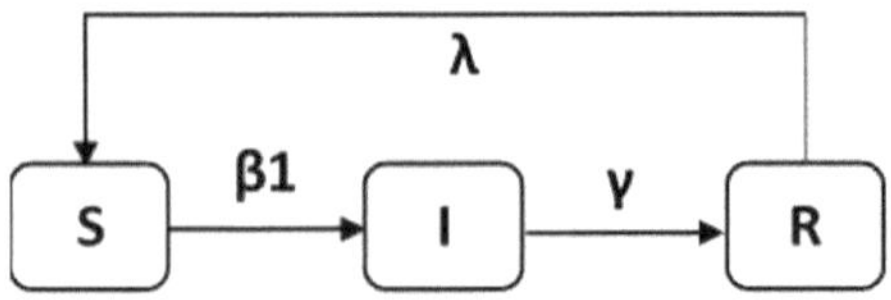

FIGURE 10.16 SIRS model

Susceptible-Infected-Recovery-Susceptible (SIRS) Model

In this model, the individuals who have recovered will lose immunity after a certain period of time and will become susceptible again.

$$\frac{dS}{dt} = \lambda R - \beta IS, \tag{10.26}$$

$$\frac{dI}{dt} = \beta IS - \gamma I, \tag{10.27}$$

$$\frac{dR}{dt} = \gamma I - \lambda R \tag{10.28}$$

Like the SIR model, this model has no closed form solution, so numerical integration can be used.

Epidemic Intervention

Suppose that we have a susceptible society and want to prevent more spread by vaccinating the most vulnerable individuals then how do we find the most vulnerable individuals? This can be done by randomly picking some nodes and asking them who is the most vulnerable from their point of view, then vaccinate those individuals.

10.4 PROBLEMS FOR SELF-ASSESSMENT

Q1. For the following graph, assume that node i activates node j when $\boldsymbol{i - j \equiv 1}$ **(mod 3)**. Solve cascade maximization for $k = 2$.

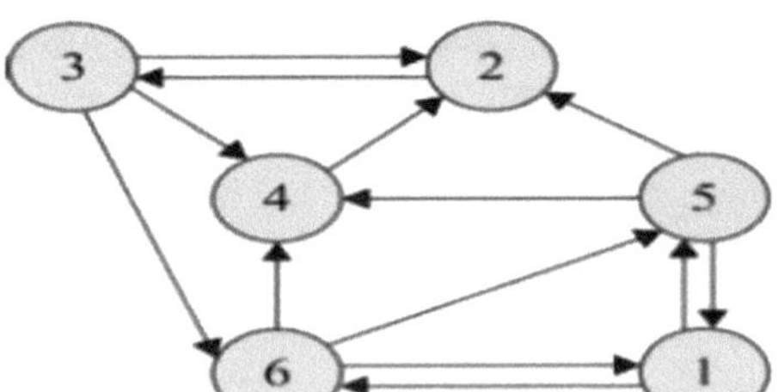

Q2. What are the minimum requirements for a herd behaviour experiment? Design an experiment of your own using conditional probability.

REFERENCES

1. Berger, J., & Milkman, K. L. (2012). What makes online content viral? *Journal of Marketing Research*, 49(2), 192–205.
2. Guille, A., Hacid, H., Favre, C., & Zighed, D. A. (2013). Information diffusion in online social networks: A survey. *ACM Sigmod Record*, 42(2), 17–28.
3. Zafarani, R., Abbasi, M. A., Liu, H., & Draft, C. (2014). Information diffusion in social media. *Social Media Mining: An Introduction*. Cambridge University Press.
4. Aggrawal, N. *Information diffusion framework for virality prediction of social media content*. PhD thesis, Jaypee Institute of Information Technology.
5. Massoud, M. R. F., Donohue, K. L., & Mccannon, C. J. (2010). Options for large-scale spread of simple high-impact interventions technical report. *Institute for Healthcare Improvement*. http://www.ihi.org/resources/Pages/Publications/OptionsforLargeScaleSpreadSimpleHighImpactInterventions.aspx
6. Hanson, K-N. Two-step flow: Audience theories. https://kieshaboa.wordpress.com/audiences/audience-theories/two-step-flow-audience-theories/
7. Aggrawal, N., Arora, A., & Anand, A. (2018). Modeling and characterizing viewers of You Tube videos. *International Journal of System Assurance Engineering and Management*, 9(2), 539–546.
8. Chakrabarti, D., Wang, Y., Wang, C., Leskovec, J., & Faloutsos, C. (2008). Epidemic thresholds in real networks. *ACM Transactions on Information and System Security (TISSEC)*, 10(4), 1–26.

11 Security and Privacy in Social Networks

11.1 INTRODUCTION

Social networking sites enhance our life; the advantages that we get when using them in a good manner are numerous. But at the same time, these sites could be a main security threat if we don't have the awareness to use them in a safe way.

The Danger of Social Networks:

Social networking sites can be a source of personal and business information leaks. They can be a malware attack vector when not used carefully. Users trust their contacts to not send bad links, not try to infect their computers and take good care of their personal data. URL shorteners are a security concern and should be taken very seriously in social networks. Social networks problems from a legal point of view include violation of a user's data protection rights, identity fraud and absence of uniform rule at the international level [1].

What could go wrong in social networking sites? Your contact's account is compromised and somebody else is using it. You added somebody to your network that you thought you knew but, in fact, you did not. You added somebody you thought was trustworthy but they turn out not to be. Insufficient use of privacy controls caused you to share data with people you never intended [2]. What data that could be mined in social networks? Social networks contain a wealth of information such as user identifiers and attributes; contact information; work, education, personal information and interests and connection and usage information.

This information is treasure for attackers and criminals. So, what will attackers do with the mined data? Underground forums sell information. Your data can be mined and stored somewhere in the dark corners of the internet waiting for a criminal to pay the right price for it. Criminals can use this information to obtain birth certificates/passports and other documents.

Most Important Data for Attacker:

Email addresses: used for spam campaigns. They can be categorized to improve the impact of the campaign like age, country and other factors so that its market price is higher than just any normal email address. They can also be a great value for phishing campaigns.

Real-life addresses: often shared in social networking sites. They can be used to collect mailing databases for advertising purposes in a similar way as described above.

Date of birth: used by different companies to confirm people's identities over the telephone. Criminals have tools to automate 'date of birth' searches in social networking sites.

DOI: 10.1201/9781003088066-11

Public profile: a potential massive data-leak age. Social networking search engines can search all available data about any name in a certain region. Public profile makes the lives of stalkers, fraudsters or any other attackers much easier.

Privacy in a Connected World – A Story:

In July 2009, the wife of a high-level government executive in the UK published personal data in a social networking site. This had got a lot of attention, not for the confidentiality of the content, but for the lack of awareness about the accessibility of your online content. There is also another issue to play here, the fact that once you publish anything online, you will lose control over it, people leech and republish it on places you do not even know.

Attackers can fool social networks users by creating a fake celebrity profile or a duplicate of somebody's profile, adding themselves to a medium-sized group or community and then joining a second group and starting again. They can also fool us by creating a female profile and publishing a pretty picture of 'herself'. Attackers [3] succeed in fooling social network users because lot of users don't realize that their contact lists are a circle of trust. And by adding somebody they don't know, they are opening their data to untrusted parties. Some sites don't have privacy controls in place, or the ones they have do not protect all user data. The user is often not concerned with selecting who can access their data. The user doesn't use the available controls because they appear too complex or time-consuming (laziness or lack of knowledge). Attackers always try to create a large network and in that large network they advertise and collect contact information, they do phishing and malware [4] installation. Sometimes, social networking sites have code breaks. Attack vector is the exploitation of programming flaws in websites. These web pages have been made by humans and they can have errors that could compromise the site's security measures. Poor security, weak administration practices or badly written code can all help attackers to gather your data or help them stage a bigger attack against any number of users. Social networking sites keep adding to their security controls and refining their existing ones. But as in any development project, they also continue to innovate on their platforms and add exciting new features.

These new options need to keep up with the security features or they too will suffer from security weaknesses.

Code Break Examples on Various Platforms:

Pinterest: in Pinterest, a cross-site scripting vulnerability and an iframe injection issue had been identified that could allow hackers to hijack user accounts and perform other malicious operations. A URL redirection flaw had been found that could redirect the site's visitors to other potentially malicious domains.

Facebook: there have been instances of security flaws on Facebook that allowed anybody to access the 'basic information' data of any user, no matter what their security settings were. This attack was released by casual users after Facebook ignored the users' warnings for a few days. No great knowledge was needed in this case to exploit a security weakness.

Twitter: Twitter had 'cross-site scripting' attacks performed against it. In these attacks, the attackers could change the Twitter status of any user accessing the attacker's account. This meant that the bad guys could make you tweet dangerous links so your Twitter followers would be at risk of being infected.

Myspace: Myspace was attacked in 2007 by a JavaScript that would copy itself to the viewer's profile along with a piece of text – 'Sammy is my hero'. This was caused by a security flaw that could have caused the victim to run any other command like redirecting the page to a malicious website. Thankfully, the young man who discovered the flaw and created the worm only wanted to have more friends added to his profile.

These four examples are not the only cases of security flaws on social networking sites. In fact, such flaws are identified frequently. News about such security holes is released every month and are a concern for all affected websites and their users. Since their solution is out of the user's hands, it is difficult or impossible to do anything about them.

There are some simple steps and points that we should consider to protect our privacy in social networks such as:

Keep your information as general as possible, read the privacy policy of social networking sites and do not add people that you don't know or trust. If you are using social networks to meet new people, you should create a special account for that. Do not click any unknown or suspicious hyperlinks [5]. Remember that any data you publish in the internet will be there forever and anyone could use and share it. Login with a strong password. Don't reveal personal information. Don't enable auto login. Turn on cookie notices in your web browser and use cookie management software. Keep a 'clean' email address. Don't reveal personal details to strangers or just-met 'friends'. Avoid sending highly personal email to mailing lists, and keep sensitive files on your home computer. Do not reply to spammers, for any reason. Be conscious of web security. Use encryption [6].

11.2 NEED OF PRIVACY

Privacy preserving data mining has become increasingly popular because it allows sharing of privacy sensitive data for analysis purposes. Suppose a hospital has some person-specific patient data that it wants to publish. It wants to publish such that:

- Information remains practically useful.
- Identity of an individual cannot be determined.

TABLE 11.1
Sensitive and Non-Sensitive Data

	Non-Sensitive Data			Sensitive Data	
#	ZIP Code	Age	Nationality	Name	Condition
1	13053	28	Indian	Kumar	Heart Disease
2	13067	29	American	Bob	Heart Disease
3	13053	35	Canadian	Ivan	Viral Infection
4	13067	36	Japanese	Umeko	Viral Infection

TABLE 11.2
Published Data

	Non-Sensitive Data			Sensitive Data
#	**ZIP Code**	**Age**	**Nationality**	**Condition**
1	13053	28	Indian	Heart Disease
2	13067	29	American	Heart Disease
3	13053	35	Canadian	Viral Infection
4	13067	36	Japanese	Viral Infection

TABLE 11.3
Sensitive and Non-Sensitive Data Leak
Data Leak

#	Name	ZIP Code	Age	Nationality
1	John	13053	28	American
2	Bob	13067	29	American
3	Chris	13053	23	American

Quasi Identifiers

Key attributes that can be uniquely identified and always removed before release. Example: name, address, phone number.

Quasi attributes are a set of attributes that could potentially identify a record owner when combined with publicly available data. In other words, we can say these attributes in combination can uniquely identify individuals such as birth date and gender. The set of such attributes has been termed. Example: ZIP Code.

Sensitive attributes are a set of attributes that contain sensitive person specific information such as disease, salary etc. This data is used by researchers and released unmodified. Example: medical records.

Non-sensitive attributes are a set of attributes that create no problem if revealed even to untrustworthy parties.

11.3 SOCIAL NETWORK PRIVACY MODEL

We consider the social network model as a simple undirected graph $G = (N, E)$, where N is the set of nodes and $E \subseteq N \times N$ is the set of edges. Each node represents an individual entity. Each edge represents a relationship between two entities.

The set of nodes, N, is described by a set of attributes that are classified into the following three categories:

- $I_1, I_2, I_3, \ldots, I_m$ are identifier attributes such as Name and SSN that can be used to identify an entity.
- $Q_1, Q_2, Q_3, \ldots, Q_q$ are quasi-identifier (QI) attributes such as ZIP Code and Sex that may be known by an intruder.

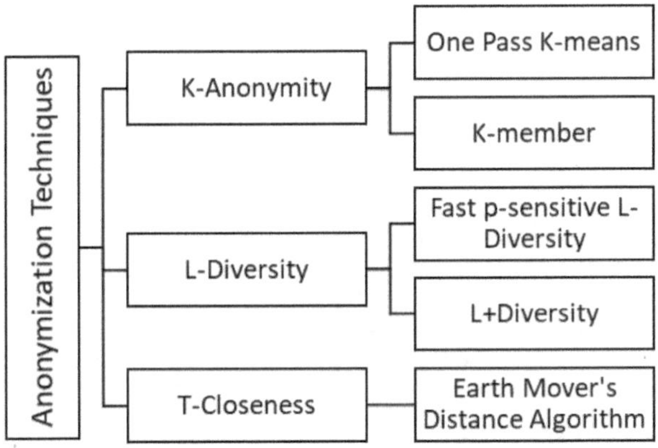

FIGURE 11.1 Classification of privacy techniques

- S_1, S_2, S_3, …, S_r are confidential or sensitive attributes (SAs) such as Diagnosis and Income that are assumed to be unknown to an intruder.

Classification of anonymization techniques: data anonymization is the process applied on the data to prevent identification of individuals, making it possible to share and analyze data securely. Figure 11.1 shows the classification of different anonymization techniques and the algorithms used by those techniques.

11.4 BASIC CONCEPTS IN DATA PRIVACY

A successful model for microdata privacy protection is *k*-anonymity, which ensures that every individual is indistinguishable with other $(k - 1)$ individuals in terms of their QI attributes' values. For social network data, the *k*-anonymity model has to impose both the QI attributes and the QI relationship homogeneity, for groups of at least *k* individuals.

The generalization of the QI attributes is one of the techniques widely used for microdata *k*-anonymization. It consists of replacing the actual value of an attribute with a less specific, more general value that is faithful to the original. We reuse this technique for the generalization of nodes attributes' values.

11.4.1 K-Anonymity

The *k*-anonymity model was developed to prevent the possibility of indirect identification of records from public databases. Combinations of record attributes can be used to exactly identify individual records. In the *k*-anonymity method [7], we reduce the granularity of data representation with the use of techniques such as generalization and suppression. The granularity is reduced sufficiently so that any given record maps onto at least *k* other records in the data.

When we discuss publishing the micro data, then we think about several types of risks associated with the data. To limit the risk of occurrence, the confidentiality of

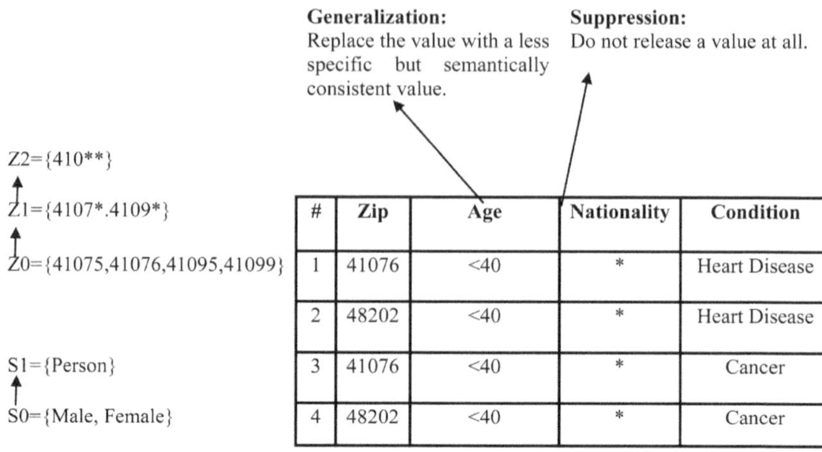

FIGURE 11.2 *k*-anonymity ways

k-anonymity is provided [8], which requires that each record in the table standard does not stand out at least ***k*** – 1 other records within the data set towards a set of QI attributes.

To achieve *k*-anonymity, there are two main ways to do it:

1. **Generalization:** replace QIs with less specific and general value within a certain range, but semantically consistent values until get *k* identical values and partition ordered-value domains into intervals. Example: 10–20
2. **Suppression:** when generalization causes too much information loss then the QI is omitted (replaced by '*'), not released at all. This is common with outliers.

There are two attacks in *k*-anonymity: a homogeneity attack and a background knowledge attack.

In a homogeneity attack, if a group member has the same set and has a similar attribute, it may be a chance to leak information. A background knowledge attack is

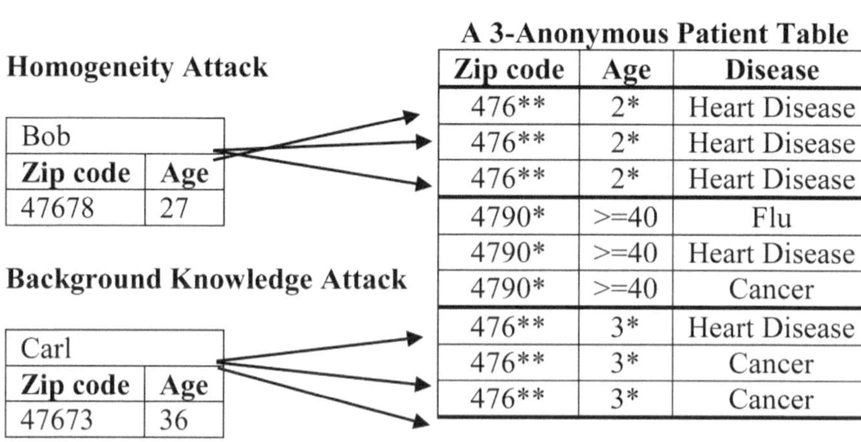

FIGURE 11.3 Attacks in *k*-anonymity

a type of attack in which if an attacker has past information about a particular record, he can easily guess about the particular records.

So, *k*-anonymity does not provide privacy if sensitive values in an equivalence class lack diversity or the attacker has background knowledge. Thus, it's a very weak privacy guarantee.

Therefore, a protected data set is said to satisfy *k*-anonymity for $k > 1$ if, for each combination of QI values (e.g., address, age, gender etc.), at least *k* records exist in the data set sharing that combination. For a given *k*, if *k*-anonymity is assumed to provide enough protection, one can concentrate on minimizing information loss with the only constraint that *k*-anonymity should be satisfied. Since *k*-anonymity is usually achieved via generalization (equivalent to global recoding) and local suppression, minimizing information loss usually translates to reducing the number and/or the magnitude of suppressions.

11.4.2 L-Diversity

The *l*-diversity was designed to handle some weaknesses in the *k*-anonymity model. Protecting identities to the level of *k* individuals is not the same as protecting the corresponding sensitive values, especially when there is homogeneity of sensitive values within a group. To handle such requirements, the concept of intra-group diversity of sensitive values is promoted within the anonymization scheme. While *k*-anonymity protects against identity disclosure, it does not provide sufficient protection against attribute disclosure. The notion of *l*-diversity attempts to solve this problem by requiring that each equivalence class has at least *l* well-represented values for each SA. The technology of *l*-diversity [9] has some advantages than *k*-anonymity, because the *k*-anonymity data set permits strong attacks due to lack of diversity in the SAs.

The *l*-diversity principle: an equivalence class is said to have *l*-diversity if there are at least *l* well-represented values for the SA. A table is said to have *l*-diversity if every equivalence class of the table has *l*-diversity. While the *l*-diversity principle represents an important step beyond *k*-anonymity in protecting against attribute disclosure, it has several shortcomings also, *l*-diversity may be difficult and unnecessary to achieve. Besides, *l*-diversity is insufficient to prevent attribute disclosure.

Attack on the *l*-diversity: when the SA values in an equivalence class are distinct but semantically similar, an adversary can learn important information. Here is an example of a similarity attack. Table 11.4 is the original table and

TABLE 11.4
Original Salary and Disease Table

Sex	ZIP Code	Salary	Disease
M	555402	K	Gastric ulcer
M	555722	2K	Gastritis
M	555801	3K	Stomach cancer
M	987789	32K	Headache
F	987056	19K	Cough
M	987444	8K	Toothache

TABLE 11.5
A 3-Diverse Version of Table 11.4

Sex	ZIP Code	Salary	Disease
M	555*	K	Gastric ulcer
M	555*	2K	Gastritis
M	555*	3K	Stomach cancer
*	987*	32K	Headache
*	987*	19K	Cough
*	987*	8K	Toothache

Table 11.5 shows an anonymous version satisfying 3-diversity. There are two SAs: Salary and Disease. Suppose one knows that Lily's record corresponds to one of the first three records, then one knows that Lily's salary is in the range [1K–3K] and can infer that Lily's salary is relatively low. This attack applies not only to numeric attributes like 'Salary', but also to categorical attributes like 'Disease'. Knowing that Lily's record belongs to the first equivalence class enables one to conclude that Lily has some stomach-related problems, because all three diseases in the class are stomach-related.

The bucketization technique of *l*-diversity can solve the problem of *k*-anonymity. In bucketization, all separate buckets are created and separate the QI and SA In *l*-diversity, two major attacks are associated: skewness attack and similarity attack. The skewness attack when overall distribution is skewed, satisfying *l*-diversity, does not prevent attribute disclosure. In such type of attack, there is a chance that anyone of a similar class can guess 50% being possibility of positive attribute linkage. In a similarity attack when the SA values in an equivalence class are distinct but semantically similar, then an adversary can learn necessary information from records. The bucketization technique has better data utility than generalization, but it has several limitations. First, bucketization does not prevent membership disclosure. Because bucketization publishes the QI values in their original forms, an adversary can find out whether an individual has a record in the published data or not. As shown, 87% of the individuals in the United States can be uniquely identified using only three

TABLE 11.6
Bucketized Data of Voter Registration List

	Sex	Age	ZIP Code	BID
(Ann)	Male	30	13053	1
(Bob)	Male	32	13053	1
(Joy)	Male	35	13053	1
(Joni)	Female	38	13054	2
(Ed)	Female	39	13054	2
(Joy)	Male	40	13054	2
(John)	Male	50	13055	3
(Rani)	Female	56	13055	3
(Salu)	Male	59	13055	3

attributes (Birth Date, Sex and ZIP Code). Micro data (e.g., census data) usually contains many other attributes besides those three attributes. This means that the membership information of most individuals can be inferred from the bucketized table.

Three buckets are created and identified by their bucket IDs (BID). In the bucketization technique, attributes are also partitioned into columns; one column contains QI values and the other column contains SA values. In bucketization, one separates the QI and SA values by randomly permuting the SA values in each bucket. In some cases, we cannot determine the difference between the two. So, it has one drawback for micro data publishing. Some limitation of *l*-diversity is as follows:

1. It does not provide a membership disclosure attack. When the attributes are less than its success but the attributes increase, such as when we analyse the census data membership information of most individuals, it can be inferred from the bucketized table.
2. Bucketization requires a clear separation between QIs and SAs. However, in many data sets, it is unclear which attributes are QIs and which are SAs.
3. When we separate the SAs from the QI attributes, bucketization breaks the attribute correlations between the QIs and the SAs.

***l*-diversity example:**

TABLE 11.7
Original Salary Disease Table

#	ZIP Code	Age	Salary	Disease
1	47677	29	3K	Gastric ulcer
2	47602	22	4K	Gastritis
3	47678	27	5K	Stomach cancer
4	47905	43	6K	Gastritis
5	47909	52	11K	Flu
6	47906	47	8K	Bronchitis
7	47605	30	7K	Bronchitis
8	47673	36	9K	Pneumonia
9	47607	32	10K	Stomach cancer

TABLE 11.8
A 3-Diverse Version of Table 11.7

#	ZIP Code	Age	Salary	Disease
1	476**	2*	3K	Gastric ulcer
2	476**	2*	4K	Gastritis
3	476**	2*	5K	Stomach cancer
4	479**	>=40	6K	Gastritis
5	479**	>=40	11K	Flu
6	479**	>=40	8K	Bronchitis
7	476**	3*	7K	Bronchitis
8	476**	3*	9K	Pneumonia
9	476**	3*	10K	Stomach cancer

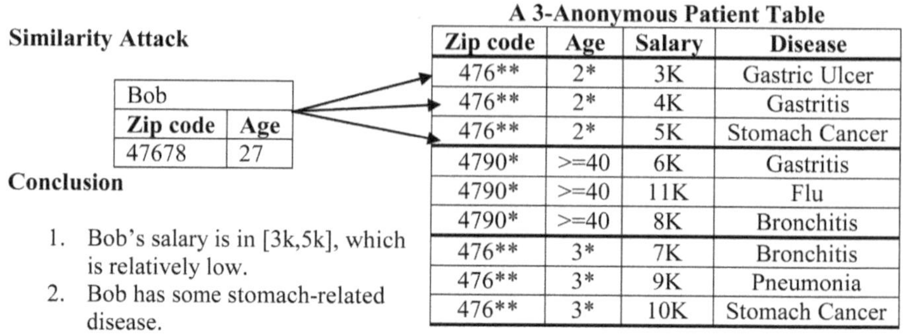

FIGURE 11.4 Attack on *l*-diversity

Limitation of *l*-diversity: *l*-diversity is insufficient to prevent attribute disclosure.

This disclosure risk of sensitive information occurs because while *l*-diversity requirement ensures diversity of sensitive values in each group, it does not take into account the semantic closeness of these attribute values (i.e., *l*-diversity does not consider semantic meanings of sensitive values).

11.4.3 T-Closeness

Sometimes, an adversary has information about statistical information that 80% of males in a particular university aged above 42 have cancer. *T*-closeness attempts to guarantee privacy against such adversaries by assuming that the distribution of the SA (say disease) in the whole table is public information. Privacy is said to be breached when the distribution of the SA in an equivalence class is not close to the distribution of the SA in the whole table.

Principle of *t*-closeness: an equivalence class is said to have *t*-closeness if the distance between the distributions of an SA in this class and the distribution of the attribute in the whole table is no more than a threshold *t*. A table is said to have *t*-closeness if all equivalence classes have *t*-closeness.

TABLE 11.9
Example of *t*-Closeness of a Medical Aspect

Non-Sensitive		Sensitive	
Age	**Gender**	**Disease**	**Count**
<42	Male	Flu	400
<42	Male	Flu	200
>42	Male	Cancer	400
>42	Female	Cancer	200
>42	Female	Cancer	400

The *t*-closeness is a guard against an adversary who knows the marginal distribution of an SA in a given record. For example, in the above table, an age less than 42 is 'Flu' and an age greater than 42 is 'Cancer'.

Limitations of *t*-closeness: some limitations of *t*-closeness also exist:

1. It lacks the flexibility of specifying different protection levels for different sensitive values.
2. The EMD function is not suitable for preventing attribute linkage on numerical SAs.
3. Enforcing *t*-closeness would greatly degrade the data utility because it requires the distribution of sensitive values to be the same in all QI groups.

Therefore, *t*-closeness is not a suitable method for privacy preservation of published data as this would significantly damage the correlation between QIs and SAs. One way to decrease the damage is to relax the requirement by adjusting the thresholds with the increased risk of skewness attack.

11.5 RANDOMIZATION

The randomization technique is used to hide data by adding noise and adding a mask with the table. The noise added is rightfully large so that the individual values of the records can no longer be recovered by any adversary [11]. In general, the randomization method aims at finding an appropriate balance between privacy preservation and knowledge discovery. Representative randomization methods include random-noise based perturbation and a randomized-response scheme [10].

11.6 SLICING

The new privacy preservation technique that is mostly used provides more security compared to the above methods. Slicing partitions the data both horizontally and vertically. Vertical partitioning is done by grouping attributes into columns based on the correlations among the attributes. It is a more secure method compared to generalization and bucketization, because all of the following steps are involved in the slicing process [10]:

Step 1: Extract the data set from the database.
Step 2: Anonymity process divides the records into two.
Step 3: Interchange the sensitive values.
Step 4: Multi-set values generated and displayed.
Step 5: Attributes are correlate and secure data displayed.

In Figure 11.5, flow of data is happening in a secure manner. Also, our slicing algorithms are being performed in the following three steps:

- Attribute partition
- Tuple partition
- Column partition

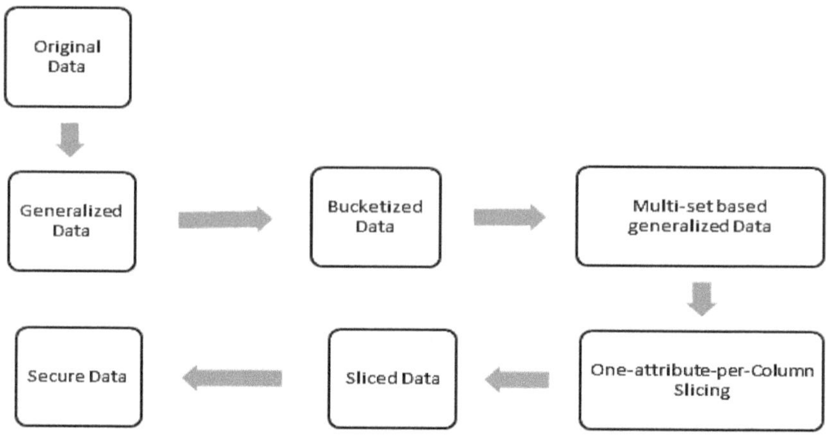

FIGURE 11.5 Slicing architecture

Slicing first partitions attributes into columns. Each column contains a subset of attributes. This vertically partitions the table. Slicing also partition tuples into buckets. Each bucket contains a subset of tuples. An attribute partition consists of several subsets of A, such that each attribute belongs to exactly one subset. Each subset of S attributes is called a column.

In attribute partitioning, attributes are highly correlated to each other in the same column, which is nice for utility and privacy. Utility means grouping highly correlated attributes and privacy means associating of uncorrelated attributes that present a high degree of identification risk and the value of the attribute is less frequent. Column generalization is not an indispensable phase in slicing algorithm as bucketization provides the same level of privacy protection, with respect to attribute disclosure; though column generalization can be useful in several aspects. First, column generalization may be required for identity/participation disclosure protection. Second, when column generalization is applied to achieve the same level of privacy against attribute disclosure, bucket sizes can be smaller. While column generalization may result in information loss, smaller bucket sizes allow better data utility. Therefore, there is a trade-off between column generalization and tuple partitioning. In tuple partitioning phase, tuples are partitioned into buckets, no generalization is applied to the tuples. If the column value is unique then it has only one matching bucket. The main problem is that a unique value is easily identified. In tuple partition, the main part is that it checks to satisfy the *l*-diversity or not. The tuple partition phase is used for data structure such as queue of bucket (Q) and sliced bucket (Sb).

Table 11.10 is truly sliced, with four rows in the upper row and four rows in the downward row. The attributes of the table are combined in such a way such as Age and Sex correlates with ZIP Code and Disease in 16 possible ways.

For example (22, M) can correlate with (14589, M), (14586, F), (14587, M) and row (14588, M). So, the partition of row and column are most secured from the outer world, because nobody can guess what exact row has the personal information.

Slicing architecture is explained in Figure 11.5, which shows several phases of data flow through step by step. The original table contains data displayed in an

TABLE 11.10
Sliced Data

(Age, Sex)	(ZIP Code, Disease)
(22, M)	(14589, M)
(22, F)	(14586, F)
(33, M)	(14587, M)
(54, M)	(14588, M)
(56, M)	(145587, M)
(60, F)	(14577, F)
(60, M)	(14788, M)
(64, M)	(14566, M)

understandable manner. In the second stage, data is generalized in a secure manner. But due to some problem, that table is bucketized and in the next stage, the data comes through various phases secured.

Thus, in this chapter, we present various approaches of anonymization for providing privacy preservation of a database. Several anonymization methods are described such as *k*-anonymity, *l*-diversity, *t*-closeness, randomization and data slicing. Data slicing overcomes the limitations of generalization and bucketization and preserves better utility while protecting against privacy threats. We illustrate how slicing is used to prevent attribute disclosures. The general methodology of this work is that before data anonymization, one can analyze the data characteristics in data anonymization. The basic idea is that one can easily design better anonymization techniques when we know the data perfectly. Finally, we have showed some advantages of data slicing comparing with generalization and bucketization. Data slicing is a promising technique for handling high-dimensional data and maintaining the high correlation of data and applying permutation and combination over the data for a privacy preservation aspect.

11.7 PROBLEMS FOR SELF-ASSESSMENT

Q1. Show the anonymized data to achieve *k*-anonymity of $k = 3$.

Name	Postcode	Age	Gender	Disease
Patrick	SW14YB	22	Male	Cardiovascular
Sebastian	SW14ZE	23	Male	Respiratory
Reece	SW12HY	18	Male	No Illness
Tilly	NW108FN	47	Female	Cancer
Abby	NW104AB	42	Female	No Illness
Elise	NW100FW	56	Female	Cardiovascular
Morgen	E179QY	23	Male	Respiratory
George	E173SF	29	Male	Liver
Sienna	E175WD	18	Female	Cancer

Q2. For inpatient micro-data, find 4-Anonymous Inpatient Microdata and 3-Diverse Inpatient Microdata:

	Non-Sensitive			Sensitive
	ZIP Code	**Age**	**Nationality**	**Condition**
1	13053	28	Russian	Heart Disease
2	13068	29	American	Heart Disease
3	13068	21	Japanese	Viral Infection
4	13053	23	American	Viral Infection
5	14853	50	Indian	Cancer
6	14853	55	Russian	Heart Disease
7	14850	47	American	Viral Infection
8	14850	49	American	Viral Infection
9	13053	31	American	Cancer
10	13053	37	Indian	Cancer
11	13068	36	Japanese	Cancer
12	13068	35	American	Cancer

REFERENCES

1. Use of social network analysis in Web 2.0. (2009). http://www.hausarbeiten.de/faecher/vorschau/147360.html
2. Threat encyclopedia. *Trend Micro*. https://www.trendmicro.com/vinfo/us/threat-encyclopedia/
3. de Linera Granda, P. A., Sanchez Mouriz, N., & Cuervas-Mons Ruiloba, S. (2008). *Legal problems of social networks*. http://www.w3.org/2008/09/msnws/papers/NETWORKS_LEGAL_PROBLEMS.PDF
4. Lewis, J. (2009, Jul 5). MI6 chief blows cover as wife's Facebook account reveals family holidays, showbiz friends and links to David Irving. *Daily Mail*. http://www.dailymail.co.uk/news/article-1197562/MI6-chief-blows-cover-wifes-Facebook-account-revealsfamily-holidays-showbiz-friends-links-David-Irving.html
5. http://www.scmagazineus.com/Facebook-bloggers-reveal-way-to-peek-at-private-profiles/article/138867/
6. Mook, N. (2005). Cross-site scripting worm hits Myspace. *betanews*. http://www.betanews.com/article/CrossSite-Scripting-Worm-Hits-MySpace/1129232391
7. Oananiculaescu. (2017, 22 Apr). k-Anonymity and cluster based methods for privacy. *GitHub*. https://elf11.github.io/2017/04/22/kanonymity.html
8. Sweeney, L. "k-anonymity: A model for protecting privacy," International Journal on Uncertainty, Fuzziness and Knowledge-based Systems, vol. 10, no. 5, pp. 557–570, 2002.
9. Machanavajjhala, A., Kifer, D., Gehrke, J., & Venkitasubramaniam M. *l*-Diversity: privacy beyond *k*-anonymity (Paper, Cornell University). https://www.cs.rochester.edu/~muthuv/ldiversity-TKDD.pdf
10. Singh, A. K., Keer, N. P., & Motwani, A. (2014). A review of privacy preservation technique. *International Journal of Computer Applications*, 90(3).

12 Social Network Analysis Tools

12.1 OVERVIEW OF SOCIAL NETWORK ANALYSIS TOOLS

These tools facilitate quantitative or qualitative analysis of social networks by describing features of a network either through numerical or visual representation.

Networks can consist of anything from families [1], project teams, classrooms, sports teams, legislatures, nation-states, disease vectors, membership on networking websites like Twitter or Facebook or even the internet. Networks can consist of direct linkages between nodes or indirect linkages based upon shared attributes, shared attendance at events or common affiliations.[2] Network features can be at the level of individual nodes, dyads, triads, ties and/or edges, or the entire network. For example, node-level features can include network phenomena such as betweenness and centrality, or individual attributes such as age, sex or income.[3] Social network analysis (SNA) software generates these features from raw network data formatted in an edge list, adjacency list or adjacency matrix (also called socio-matrix), often combined with (individual/node-level) attribute data.[4] Although the majority of network analysis software uses a plain text ASCII data format, some software packages contain the capability to utilize relational databases to import and/or store network features.

Features

Visual representations of social networks are important to understand network data and convey the result of the analysis.[5] Visualization often also facilitates qualitative interpretation of network data. With respect to visualization, network analysis tools are used to change the layout, colours, size and other properties of the network representation. Some SNA software can perform predictive analysis.[6] This includes using network phenomena such as a tie to predict individual level outcomes (often called peer influence or contagion modelling), using individual-level phenomena to predict network outcomes such as the formation of a tie/edge (often called homophily models [7]) or a particular type of triad, or using network phenomena to predict other network phenomena, such as using a triad formation at time 0 to predict tie formation at time 1.

12.2 VARIOUS TOOLS

12.2.1 Gephi (Visualization and Basic Network Metrics)

Gephi is an open-source network analysis and visualization software package written in Java on the NetBeans platform. Gephi [8] is an interactive visualization and exploration platform for all kinds of networks and complex systems, and dynamic and hierarchical graphs. It is a tool for people that have to explore and understand graphs. The user interacts with the representation, and manipulates the structures, shapes and colours to reveal hidden properties. It uses a 3D render engine to display

DOI: 10.1201/9781003088066-12

large networks in real time and to speed up the exploration. A flexible and multi-task architecture brings new possibilities to work with complex data sets and produce valuable visual results.

Social media data has helped to drive network visualization to new levels of relevance and importance. However, there is far more to network visualization than just social media data. For analyzing and visualizing network graphs, you need to have an excellent platform, and you need to know ways to use your data effectively.

Main functionality: Graph exploration and manipulation software
Input format: GraphViz(.dot), Graphlet(.gml), GUESS(.gdf), LEDA(.gml), NetworkX(.graphml, .net), NodeXL(.graphml, .net), Pajek(.net, .gml), Sonivis(.graphml), Tulip(.tlp, .dot), UCINET(.dl), yEd(.gml), Gephi (.gexf), Edge list(.csv), databases
Output Format: GUESS(.gdf), Gephi(.gexf), .svg, .png
Platform: Any system supporting Java 1.6 and OpenGL
License and cost: Open source (GPL3), seeking contributors

12.2.2 NetLogo (Modelling Network Dynamics)

NetLogo can be used to teach programming, computational thinking, simulation and model building, and understanding of complex phenomena through models in many different domains. NetLogo is a cross-platform multi-agent programmable modelling environment.

Main functionality: NetLogo is a programming language and integrated development environment (IDE) for agent-based modelling.
Input format: .nlogo, .nlogo3d, .nls

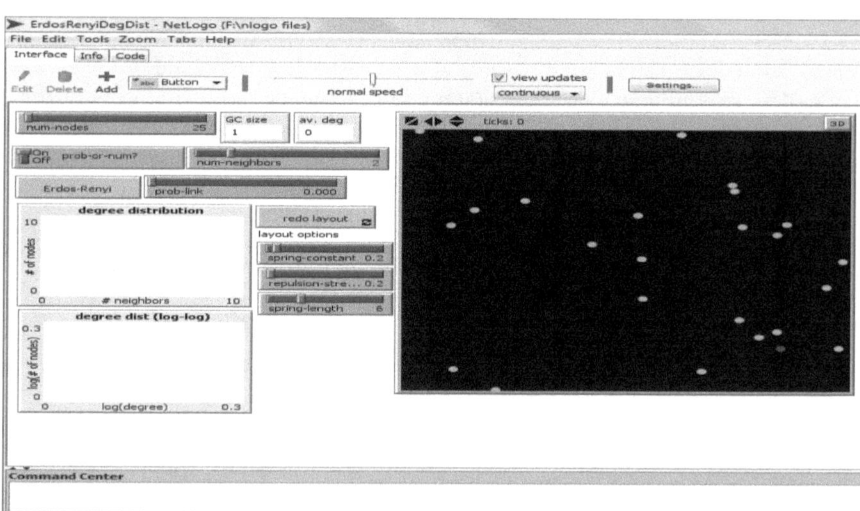

FIGURE 12.1 Erdös–Rényi model with 25 nodes and 0 probability on NetLogo tool

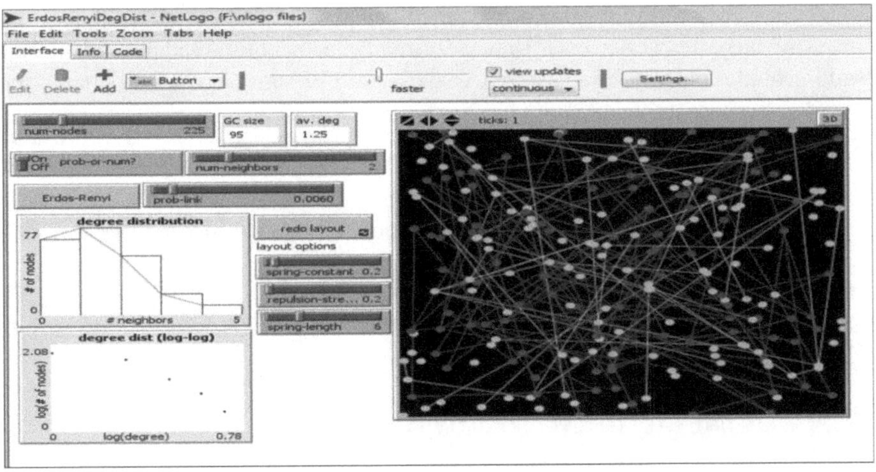

FIGURE 12.2 Erdös–Rényi model with 225 nodes and 0.0060 probability on NetLogo tool

Output format: .nlogo, .nlogo3d, .nls
Platform: IA-32, x86-64
License and cost: GPL, opensource

We used NetLogo as a modelling tool to run basic models. We tried all random models on this tool and Figures 12.1 and 12.2 show the Erdös–Rényi model.

Number of Nodes = 25 Probability = 0
Number of Nodes = 225 Probability = 0.0060
Output: Average degree: 1.25
The degree distribution graph is shown in Figure 12.2.

12.2.3 IGRAPH (FOR PROGRAMMING ASSIGNMENT)

igraph is a library collection for creating and manipulating graphs and analyzing networks. It is written in C and also exists as Python and R packages; an interface for Mathematica is also provided.

Main functionality: Efficiently create, manipulate, analyze and plot graphs in C, Python or R
Input format: GraphML, LEDA (.gml) and more
Output format: GraphML, LEDA (.gml), GraphViz (.dot) and multiple image formats
Platform: Linux, Mac, Windows
License and cost: Open source (GPL2)

12.2.4 PAJEK (USER FRIENDLY, FREE, WINDOWS ONLY)

Pajek is a program for Windows, for analysis and visualization of large networks having some thousands or even millions of vertices. In the Slovenian language the word *pajek* means spider. The latest version of Pajek is freely available for noncommercial use.

The main motivation for development of Pajek was the observation that several sources of large networks exist that are already in machine-readable form. Pajek should provide tools for analysis and visualization of such networks: collaboration networks, organic molecule in chemistry, protein-receptor interaction networks, genealogies, internet networks, citation networks, diffusion (AIDS, news, innovations) networks, data-mining (2-mode networks) etc.

Main functionality: very extensive functionality via drop-down menus
Input format: .paj,.clu,.net
Output format: .net
Platform: Windows only
License and cost: free

12.2.5 UCINET (Extensive, Socially Focused Functionality, Windows Only)

UCINET is a comprehensive package for the analysis of social network data as well as other 1-mode and 2-mode data. Can read and write a multitude of differently formatted text files, as well as Excel files. Can handle a maximum of 32,767 nodes (with some exceptions), although practically speaking many procedures get too slow around 5,000–10,000 nodes. SNA methods include centrality measures, subgroup identification, role analysis, elementary graph theory and permutation-based statistical analysis. In addition, the package has strong matrix analysis routines, such as matrix algebra and multivariate statistics.

Main functionality: Integrated with UCINET is the NetDraw program for drawing diagrams of social networks. In addition, the program can export data to Mage and Pajek.
Input format: DL format, .##h
Output format: .##h
Platform: Windows
License and cost: UCINET can be downloaded and used free for 60 days.

12.2.6 Network Overview Discovery Exploration for Excel (NodeXL) (SNA Integrated to Excel, Windows Only, Free, Beta)

NodeXL is a free and open Excel 2007 or above version Add-in and C#/. Net library for network analysis and visualization. It integrates into Excel and adds a directed graph as a chart type to the spreadsheet and calculates a core set of network metrics and scores. It supports extracting email, Twitter, YouTube, Facebook, WWW, Wiki and Flickr social networks. It accepts edge lists and matrix representations of graphs. It allows for easy and automated manipulation and filtering of underlying data in spreadsheet format. It has multiple network visualization layouts. It reads and writes Pajek, UCINET and GraphML files.

Main functionality: Network overview, discovery and exploration
Input format: email, .csv (text), .txt, .xls (Excel), .xslt (Excel 2007 or above version), .net (Pajek), .dl (UCINET), GraphML

Output format: .csv (text), .txt, .xls (Excel), .xslt (Excel 2007 or above version), .dl (UCINET), GraphML
Platform: Windows
License and cost: Free (Ms-PL)

12.2.7 NetMiner 4

NetMiner is a software tool for exploratory analysis and visualization of large network data. NetMiner 4 embeds an internal Python-based script engine that is equipped with the automatic Script Generator for unskilled users. Then, the users can operate NetMiner 4 with existing GUI or programmable script language. Analysis of large networks (+10,000,000 nodes), comprehensive network measures and models.

Both exploratory and confirmatory analysis
Interactive visual analytics
What-if network analysis
Built-in statistical procedures and charts
Full documentation (1,000+ pages of User's Manual)
Expressive network data model
Facilities for data and workflow management
Python-based Script workbench and user-friendliness
Morphological Analyzer for semantic network analysis
Main functionality: All-in-one software for network analysis and visualization
Input format: .xls(Excel),.xlsx (Excel 2007), .csv(text), .dl(UCINET), .net(Pajek), .dat(StOCNET), .gml; NMF(proprietary)
Output format: .xls(Excel),.xlsx (Excel 2007), .csv(text), .dl(UCINET), .net(Pajek), .dat(StOCNET), NMF(proprietary)
Platform: Microsoft Windows
License and cost: Free (Coursework) and Commercial

12.2.8 NetworkX (Extensive Functionality, Scales to Large Networks by Taking Advantage of Existing C, Fortran Libraries for Large Matrix Computations, Open Source)

NetworkX (NX) is a toolset for graph creation, manipulation, analysis and visualization. User interface is through scripting/command-line provided by Python. NX includes several algorithms, metrics and graph generators. Visualization is provided through Pylab and Graphviz. NX is an open-source project, in active development since 2004 with an open bug-tracking site and user forums. Development is sponsored by Los Alamos National Lab.

Main functionality: Python package for the creation, manipulation and study of the structure, dynamics and functions of complex networks.
Input format: GML, Graph6/Sparse6, GraphML, GraphViz (.dot), NetworkX (.yaml, adjacency lists and edge lists), Pajek (.net), LEDA

Output format: GML, Gnome Dia, Graph6/Sparse6, GraphML, Graphviz (.dot), NetworkX (.yaml, adjacency lists and edge lists), Pajek (.net) and assorted image formats (.jpg, .png, .ps, .svg, et al.)
Platform: Open source (GPL and similar)
License and cost: Free

12.2.9 R (Extensive, Statistics-Heavy Functionality)

R contains several packages relevant for social network analysis:
igraph is a generic network analysis package;
SNA performs sociometric analysis of networks;
network manipulates and displays network objects;
PAFit can analyse the evolution of complex networks by estimating preferential attachment and node fitness;
tnet performs analysis of weighted networks, two-mode networks and longitudinal networks;
ergm is a set of tools to analyze and simulate networks based on exponential random graph models exponential random graph models;
Bergm provides tools for Bayesian analysis for exponential random graph models;
hergm implements hierarchical exponential random graph models;
RSiena allows the analyses of the evolution of social networks using dynamic actor-oriented models;
latentnet has functions for network latent position and cluster models;
degreenet provides tools for statistical modelling of network degree distributions;
networksis provides tools for simulating bipartite networks with fixed marginals;
multiplex offers tools for the analysis of multiple social networks with algebra;
netdiffuseR was designed for the analysis of network diffusion of innovations (and diffusion in general); and
bipartite provides functions to visualize and calculate indices used to describe bipartite graphs. It focuses on webs (i.e., ecological networks).
Main functionality: Social network analysis within the versatile and popular R environment
Input format: R will read in almost any format data file
Output format: R has write capability for most data formats
Platform: Windows, Linux, Mac
License and cost: Open source

12.2.10 SocioViz

SocioViz is a social media analytics platform for digital journalists, social researchers and media marketers. It lets you analyze any topic, term or hashtag, identify key influencers, opinions and contents and export the data in Gephi format for further analysis. SocioViz is a web-based software for network analysis and visualization. It supports Twitter searches for any keyword, hashtag, emoji or user mention or Facebook analysis of public pages. Twitter and Facebook posts are represented as networks of user interactions, words and hashtags copresence. Graphs can be visualized and manipulated online or exported for further analysis.

Main functionality: Graph discovery and analysis
Input format: Public data from Twitter and Facebook pages
Output format: CSV, TXT, XLSX, GEXF (Gephi) and GML (Tulip)
Platform: Any (web based)
License and cost: Commercial (Freemium)

12.2.11 UNISoN (Social Network Analysis Tool)

UNISoN is a Java application that can download Usenet messages from free NNTP servers, show the saved messages then allow filtering of data to save to a Pajek network file or CSV file. It creates networks using the author of each post. If someone replies to a post, there is a unidirectional link created from the author of the post to the author of the message they are replying to. There is also a preview panel that shows the network visually.

Main functionality: Download usenet messages and save SNA output files
Input format: Reads from free NNTP servers
Output format: Creates CSV files and Pajek .net files
Platform: Any system supporting Java
License and cost: Open source

12.2.12 Wolfram Alpha

Wolfram Alpha is a general computational knowledge engine answering queries on many knowledge domains. Give it the input 'Facebook report' and it will answer queries on analysis of your social network data. [9, 10]

Main functionality: Graph analysis, time series analysis, categorical data analysis
Input format: Facebook API
Output format: Many formats
Platform: Web service
License and cost: Free

12.3 PROBLEMS FOR SELF-ASSESSMENT

Q1. Use the Gephi tool to analyze and visualize real-world networks. Gephi is a complete stand-alone application (available at: http://gephi.org) that you can download and install on your computer.
You can download the beta version directly from the following link: https://github.com/gephi/gephi/releases/download/v0.8.2/gephi-0.8.2-beta.setup.exe

Some data sets are available in the Data Set folder in .gml format (Graph Modelling Language)

FYI: A brief description of each of the above networks is given at the end of the question.

Use 'Fruchterman Reingold' as the layout algorithm.

Network metrics: You will determine the following and show the appropriate visualization:

- Degree distribution (node degree vs probability of node degree)
- A suitable network layout depicting the tradeoff and/or correlation between node degree and local cluster coefficient
- A suitable network layout depicting the tradeoff and/or correlation between closeness centrality and betweenness centrality values
- A suitable network layout depicting the different communities of nodes in your network and the eigenvector centrality values of the nodes.
- What is your average path length, network diameter and modularity score?

Data set description:

Zachary's karate club: social network of friendships between 34 members of a karate club at a US university in the 1970s. Please cite W. W. Zachary, An information flow model for conflict and fission in small groups, *Journal of Anthropological Research* 33, 452-473 (1977).

Les Misérables: co-appearance network of characters in the novel Les Misérables. Please cite D. E. Knuth, *The Stanford GraphBase: A Platform for Combinatorial Computing,* Addison-Wesley, Reading, MA (1993).

Word adjacencies: adjacency network of common adjectives and nouns in the novel David Copperfield by Charles Dickens. Please cite M. E. J. Newman, *Phys. Rev.* E 74, 036104 (2006).

American College football: network of American football games between Division IA colleges during regular season Fall 2000. Please cite M. Girvan and M. E. J. Newman, *Proc. Natl. Acad. Sci.* USA 99, 7821-7826 (2002).

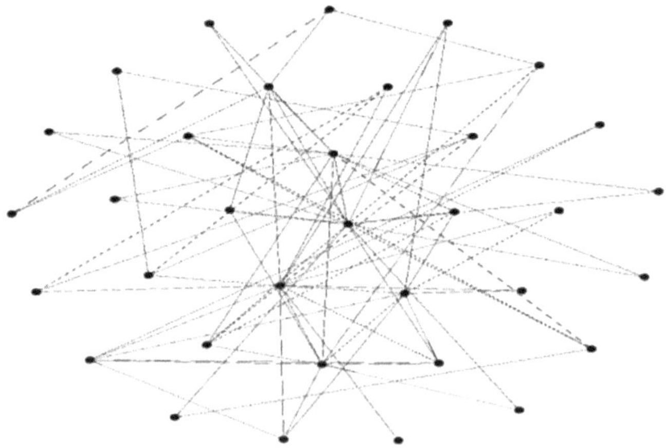

FIGURE 12.3 Visualization of Zachary's karate club

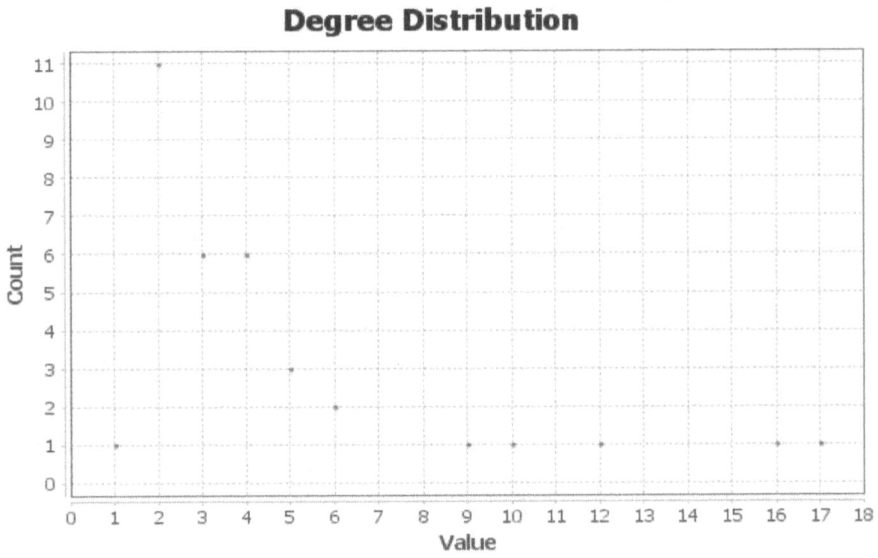

FIGURE 12.4 Degree distribution of Zachary's karate club

Source: http://konect.cc/networks/

Answer:

The data set used is **Zachary's karate club**. It is a social network of friendships between 34 members of a karate club at a US university in the 1970s. (http://konect.cc/networks/ucidata-zachary/)

Nodes: 34

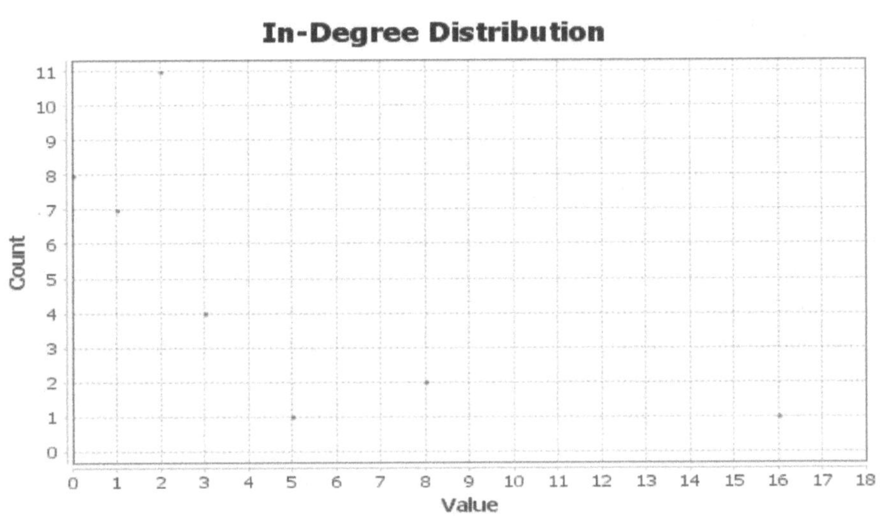

FIGURE 12.5 Indegree distribution of Zachary's karate club

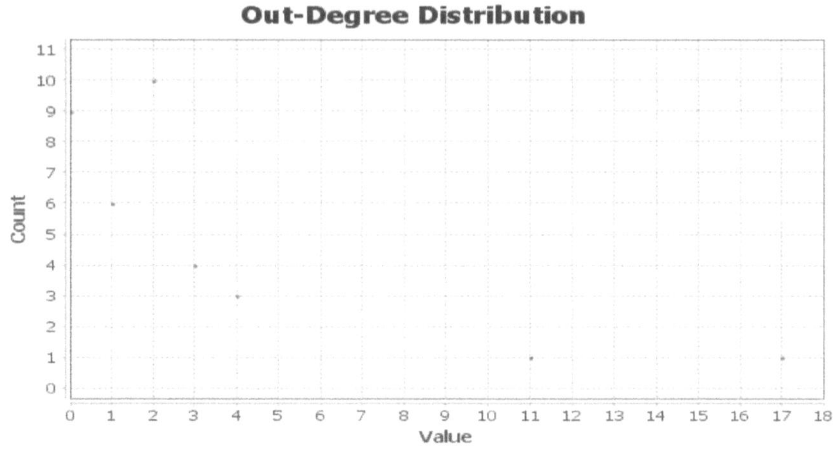

FIGURE 12.6 Outdegree distribution of Zachary's karate club

Edges: 78

We have used the **Fruchterman Reingold** layout to display the network. The solutions given are produced through the Gephi tool. After running the abovementioned layout, the network looked like the graph below.

We need to determine the following and show the appropriate visualization:

1. **Degree distribution (node degree vs probability of node degree):** To calculate the node degree, we ran the average degree functionality in the Gephi tool and got the following results:

Average Degree: 2.294

To calculate **probability of the degrees**, we have collected data from the Data Laboratory section of the Gephi tool. We used Microsoft Excel to calculate probabilities of different degrees present. Then, we plotted the graph in Figure 12.7 between

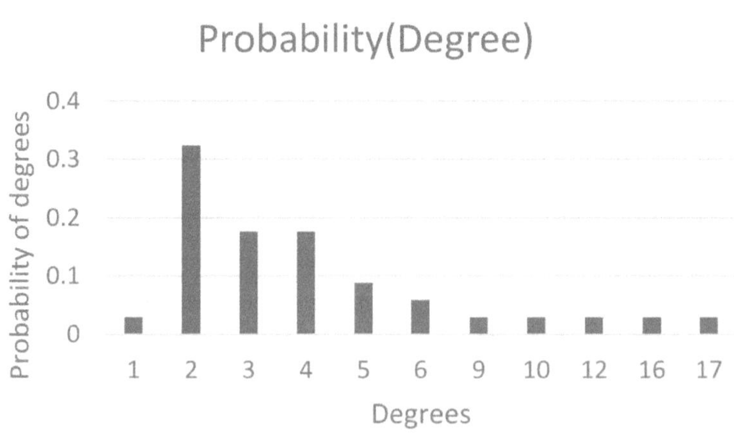

FIGURE 12.7 Probability distribution of Zachary's karate club

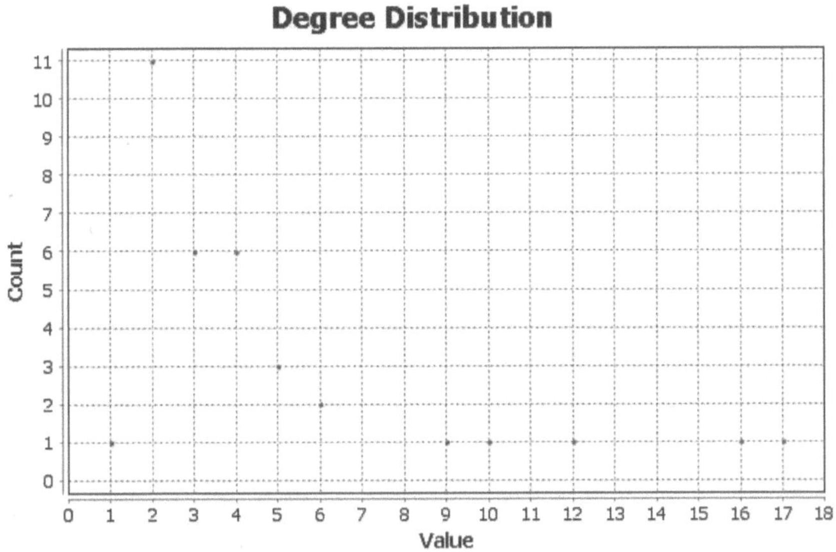

FIGURE 12.8 Correlation between node degree and local cluster coefficient of Zachary's karate club

the degrees and their probabilities, which tells us the probability of a node having that degree. We can see that the nodes with degree 2 have higher probability of occurring in the network than the other nodes.

2. **A suitable network layout depicting the trade-off and/or correlation between node degree and local cluster coefficient.**

The following is the result produced while running the clustering coefficient and average degree in the Gephi tool:

Average Degree: 2.294

The following is the result produced after running Avg. Clustering Coefficient. It tells us what portion of a node's neighbours are connected.

Average Clustering Coefficient: 0.588

Total triangles: 45

The Average Clustering Coefficient is the mean value of individual coefficients.

We have tried to find the relation between degrees and clustering coefficients of the nodes. The bigger **size** of a node indicates a larger degree, while the darker the green **colour** of the nodes, the greater are their clustering coefficients. After observing the graph, we can say smaller nodes (with lesser degree) in this graph have higher clustering coefficients. Or, we can say that degrees and clustering coefficients of the nodes are **negatively correlated**.

Size: Degree

Colour: Clustering coefficient

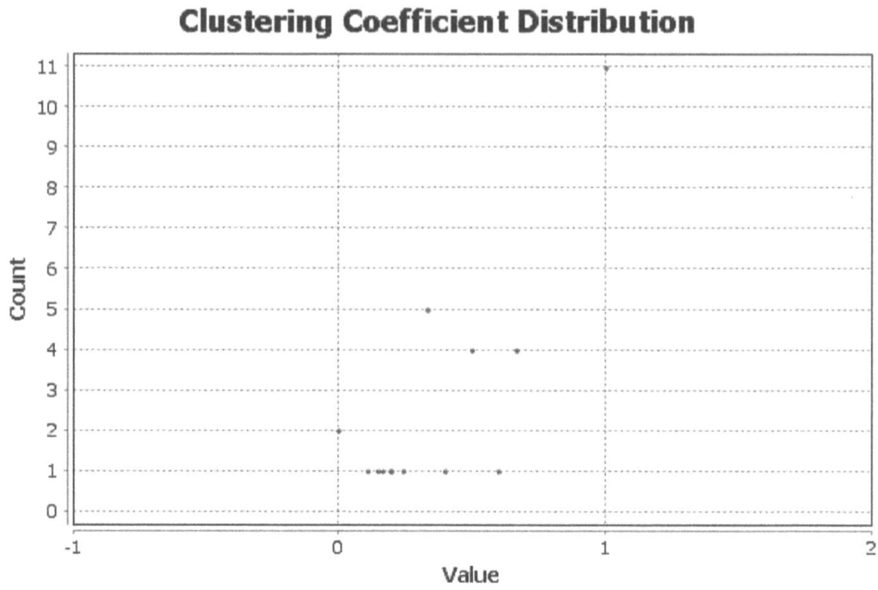

FIGURE 12.9 Average clustering coefficient of Zachary's karate club

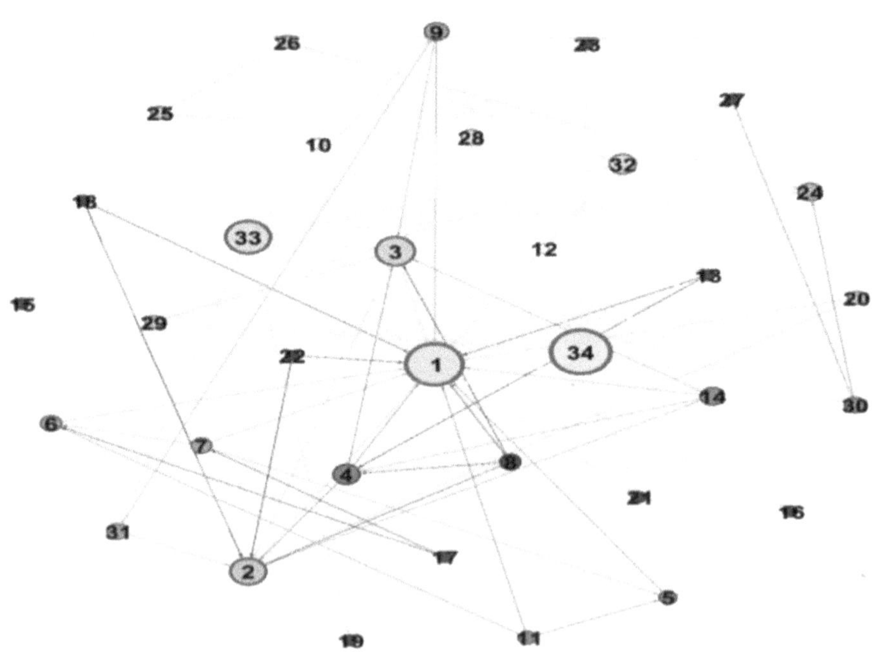

FIGURE 12.10 Correlation between degree and clustering coefficient of Zachary's karate club

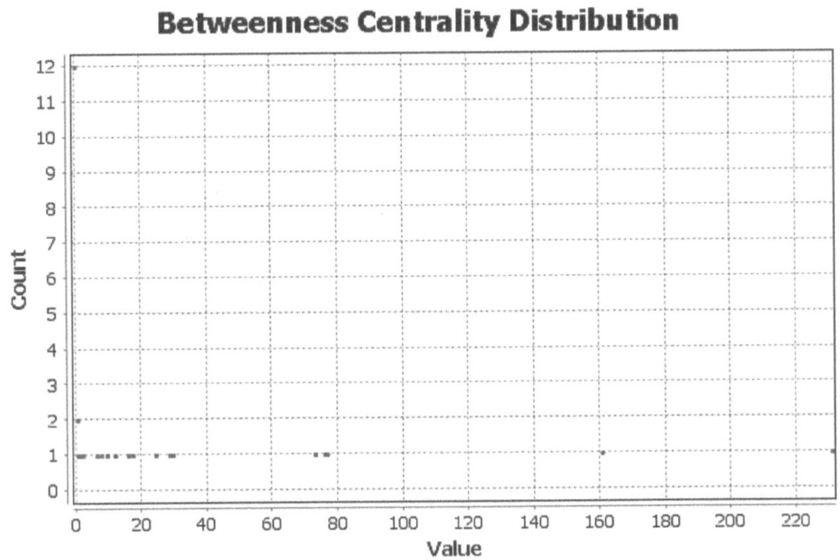

FIGURE 12.11 Betweenness centrality of Zachary's karate club

3. **A suitable network layout depicting the tradeoff and/or correlation between closeness centrality and betweenness centrality values.**

The following is the result produced while running the network diameter in the Gephi tool:

Closeness centrality tells us how fast a node can reach everyone in the network. **Betweenness centrality** tells us how likely a node to be the most direct route between two people in the network.

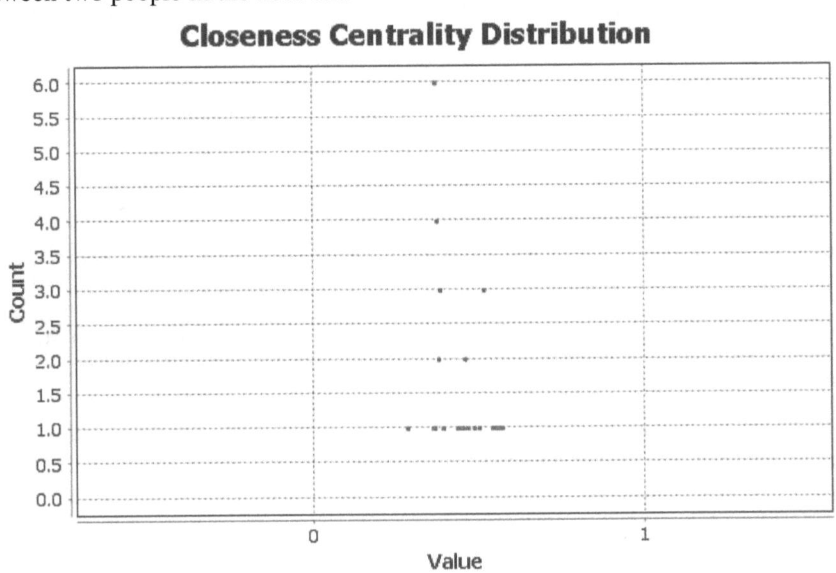

FIGURE 12.12 Closeness centrality of Zachary's karate club

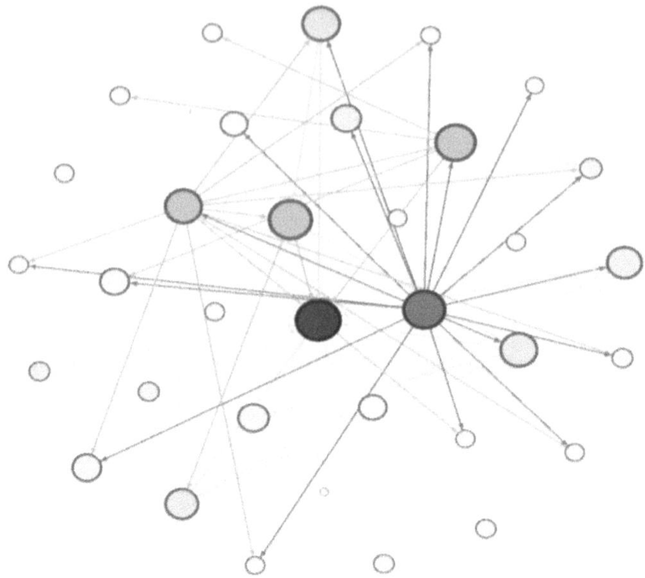

FIGURE 12.13 Correlation between closeness and betweenness centrality of Zachary's karate club

For this problem, we have tried to find relation (or correlation) between closeness centrality and betweenness centrality. We have indicated the **closeness centrality** with the **size** of the nodes, that is, larger the size of the node, greater is its closeness centrality. Also, we have tried to indicate **betweenness centrality** with the **colour** of the nodes. The darker the green colour, the greater is the betweenness centrality. And hence, it is evident from Figure 12.13 that there is **highly positive correlation** between closeness centrality and betweenness centrality, as bigger nodes are in a dark green colour.

Size: closeness centrality

Colour: betweenness centrality

4. **A suitable network layout depicting the different communities of nodes in your network and the eigenvector centrality values of the nodes.**

Selecting resolution of 1.0, I have run the modularity in the Gephi tool and obtained the following results. We can see that we have obtained three communities in our network with the specified resolution.

Modularity: 0.395

Modularity with resolution: 0.395

Number of Communities: 3

In the Figure 12.15, we have tried to show the communities present in the network with different colours. With resolution 1.0, we have **three communities** (Blue, Red, Green) in the network.

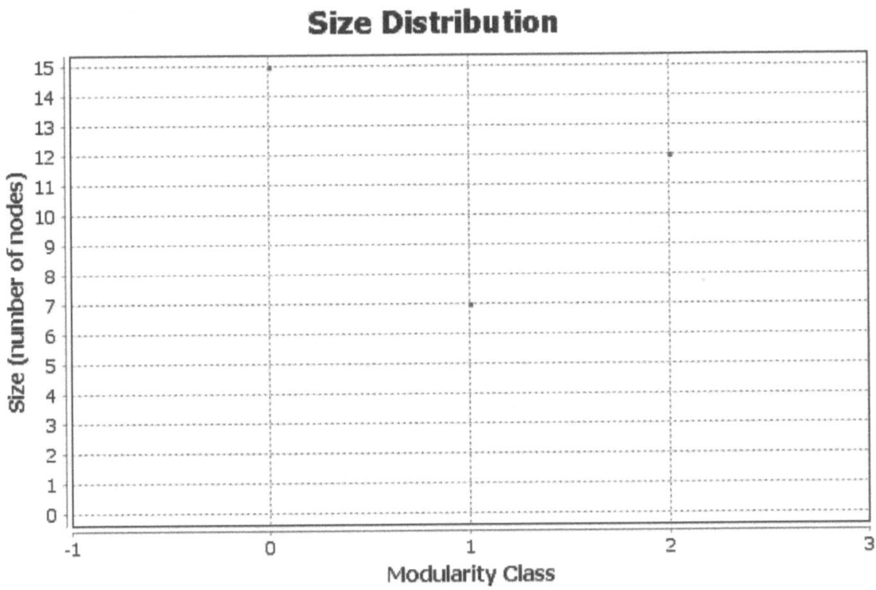

FIGURE 12.14 Community detection using modularity of Zachary's karate club

After running the eigenvector centrality, we have obtained the following results from Gephi:

Network interpretation: undirected
Number of iterations: 100
Sum change: 9.783386901414531E-4

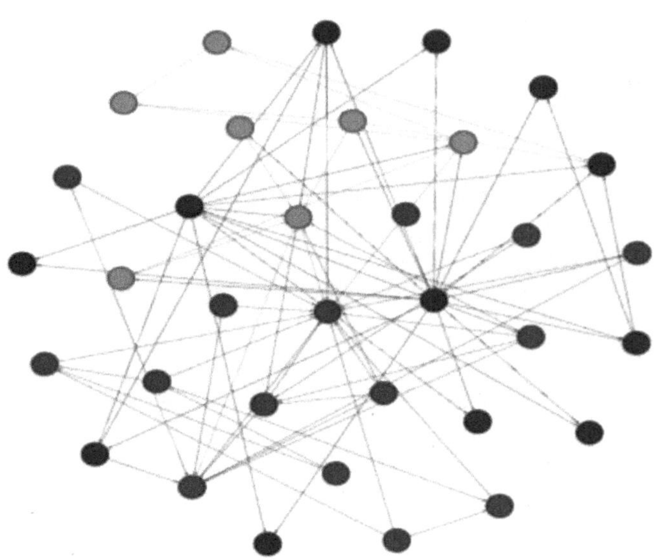

FIGURE 12.15 Community detection using modularity of Zachary's karate club

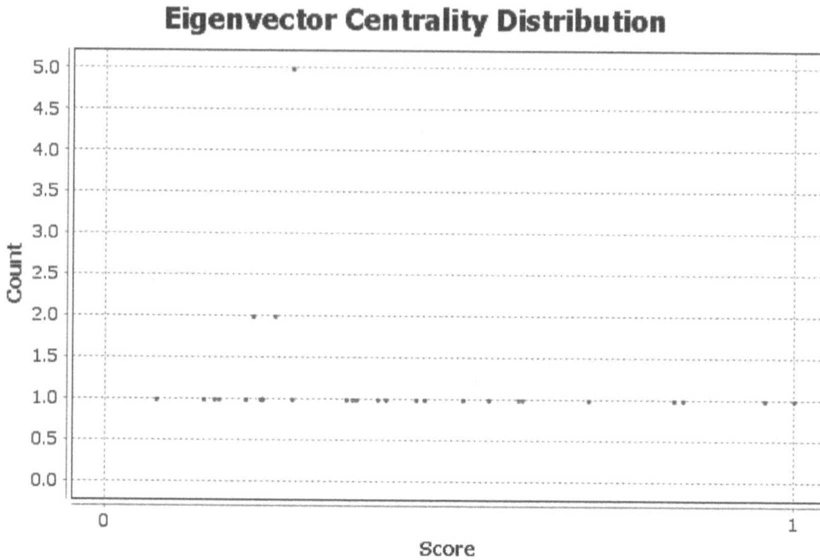

FIGURE 12.16: Eigenvector centrality of Zachary's karate club

Figure 12.17 is the graph for **eigenvector centrality**. The **darker** the green **colour**, the **higher** the eigenvector centrality of a node. This measure tells us how well a person is connected to other well-connected people.

We can see nodes 1, 3, 33 and 34 are well connected to other well-connected nodes because they are darker in colour. We can also see node 2 and node 9, which

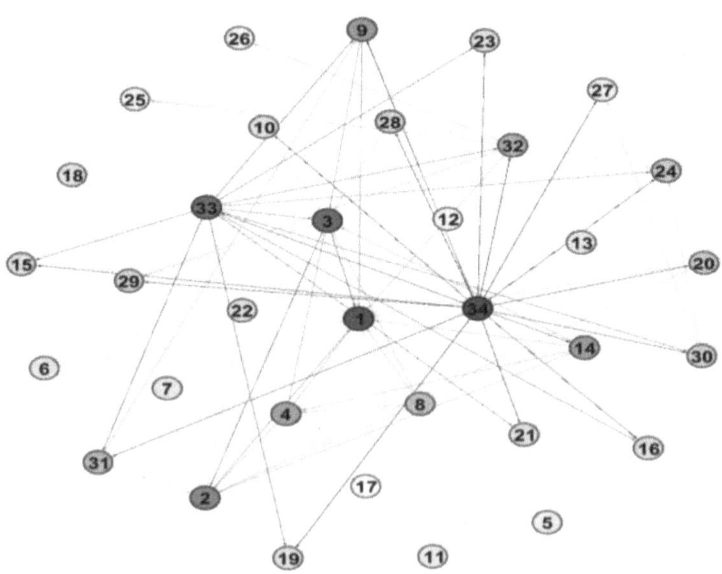

FIGURE 12.17 Eigenvector centrality of Zachary's karate club

are present on the outer area, are also well-connected to other well-connected nodes (or people) in comparison to many nodes.

5. **What is your average path length, network diameter and modularity scores?**

We have computed the below statistics of the graph from the Gephi tool:

Average path length: 2.408

This means that the average number of steps along the shortest paths for all possible pairs of network nodes is 2.408.

Network diameter: 5

In other words, the maximum (shortest path) distance between any pair of nodes is 5.

Modularity score: 0.395

This is the strength of the partition or division of the communities.

Q2. Given a Twitter data set that consists of the following two files, follow the steps to explore and visualize this data on Gephi:

EdgesTwitter.csv: a random selection of Twitter users and their 'followings' relations. The 'Edges' file contains a list of identifiers couples showing who follows who.

NodesTwitter.csv: this 'Nodes' file contains the identifiers of each node such as their label, a sex attribute and a random value that will be useful to play with visualization tools hereafter.

Step 1: Create a new project in Gephi

Run the application on your computer and create a 'new project' in the start menu. In the Data Laboratory, click on 'Import Spreadsheet' to open the import window and import your 'nodes' file.

Step 2: Import the data into Gephi

Import node data: Specify that the separation between your data is expressed by a semicolon and at the time of import in Gephi mention that import data is related to nodes.

Import edge data: Import 'edges' file and follow the same procedure as for the nodes. Specify the semicolon and import in Gephi as edges. Uncheck 'create missing nodes'.

Step 3: Visualization of data in Gephi

Now use Ranking and the Layout panel in overview to explore visualization options in Gephi.

Answer the following questions about this graph:

1. How many nodes are in the network?
2. How many edges are in the network?
3. Is this graph directed or undirected?
4. Create an adjacency matrix for this graph.
5. What is the length of the shortest path from node Alice to node Chuck?

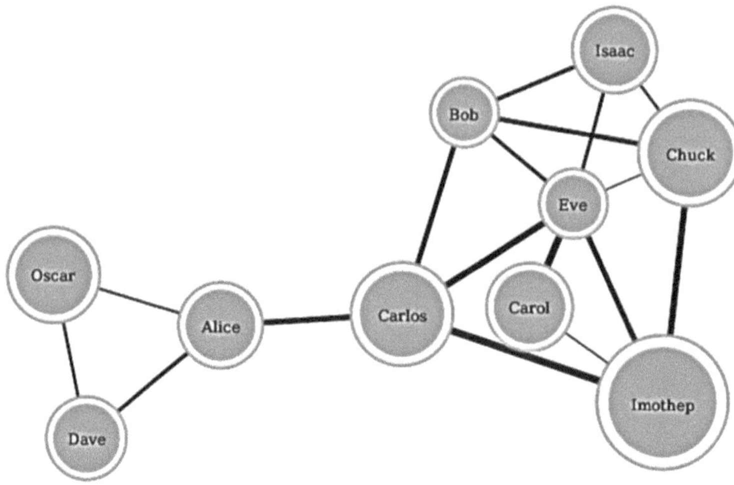

FIGURE 12.18 Example graph for Q2

Q3. Do hands-on on NetworkX using the following commands.

NetworkX is a Python language software package for the creation, manipulation and study of the structure, dynamics and function of complex networks. NetworkX is used by developers to process and perform functionalities on various social media applications. With NetworkX, you can load and store networks in standard and non-standard data formats, generate many types of random and classic networks, analyze network structure, build network models, design new network algorithms, draw networks and much more [11].

You need to install two Python packages using pip install, Packages are: NetworkX, Matplotlib (take .whl file for this from http://konect.cc/networks/)

Graph Creation

NetworkX graph objects can be created in one of three ways:

- Graph generators – standard algorithms to create network topologies.
- Importing data from pre-existing (usually file) sources.
- Adding edges and nodes explicitly.

Explicit addition and removal of nodes/edges is the easiest to describe. Each graph object supplies methods to manipulate the graph. For example:

```
>>> import networkx as nx
>>> G=nx.Graph()
>>> G.add_edge(1,2) # default edge data=1
>>> G.add_edge(2,3, weight=0.9) # specify edge data
```

Edge attributes can be anything:

```
>>> import math
>>> G.add_edge('y','x',function=math.cos)
>>> G.add_node(math.cos) # any hashable can be a node
```

You can add many edges at one time:

```
>>> elist=[('a','b',5.0),('b','c',3.0),('a','c',1.0),('c',
'd',7.3)]
>>> G.add_weighted_edges_from(elist)
```

Start here to begin working with NetworkX.

Creating a graph

Create an empty graph with no nodes and no edges,

```
>>> import networkx as nx
>>> G=nx.Graph()
```

By definition, a graph is a collection of nodes (vertices) along with identified pairs of nodes (called edges, links etc.). In NetworkX, nodes can be any hashable object (e.g. a text string, an image, an XML object, another Graph, a customized node object etc.). (Note: Python's None object should not be used as a node because it determines whether optional function arguments have been assigned in many functions.)

Nodes

The graph G can be grown in several ways. NetworkX includes many graph generator functions and facilities to read and write graphs in many formats. To get started, we'll look at simple manipulations. You can add one node at a time:

```
>>> G.add_node(1)
```

add a list of nodes,

```
>>> G.add_nodes_from([2, 3])
```

or add any nbunch of nodes. An nbunch is any iterable container of nodes that is not itself a node in the graph. (e.g., a list, set, graph, file etc.)

```
>>> H=nx.path_graph(10)
>>> G.add_nodes_from(H)
```

Note that G now contains the nodes of H as nodes of G. In contrast, you could use the graph H as a node in G.

```
>>> G.add_node(H)
```

The graph G now contains H as a node. This flexibility is very powerful because it allows graphs of graphs, graphs of files, graphs of functions and much more. It is worth thinking about how to structure your application so that the nodes are useful entities. Of course, you can always use a unique identifier in G and have a separate dictionary keyed by identifier to the node information if you prefer. (Note: You should not change the node object if the hash depends on its contents.)

Edges

G can also be grown by adding one edge at a time,

```
>>> G.add_edge(1,2)
>>> e=(2,3)
>>> G.add_edge(*e) # unpack edge tuple*
```

by adding a list of edges,

```
>>> G.add_edges_from([(1,2),(1,3)])
```

or by adding any ebunch of edges. An ebunch is any iterable container of edge-tuples. An edge-tuple can be a 2-tuple of nodes or a 3-tuple with 2 nodes followed by an edge attribute dictionary, for example, (2,3,{'weight':3.1415}). Edge attributes are discussed further below.

```
>>> G.add_edges_from(H.edges())
```

One can demolish the graph in a similar fashion; using Graph.remove_node(),Graph.remove_nodes_from(), Graph.remove_edge() and Graph.remove_edges_from(), e.g.

```
>>> G.remove_node(H)
```

There are no complaints when adding existing nodes or edges. For example, after removing all nodes and edges,

```
>>> G.clear()
```

we add new nodes/edges and NetworkX quietly ignores any that are already present.

```
>>> G.add_edges_from([(1,2),(1,3)])
>>> G.add_node(1)
>>> G.add_edge(1,2)
>>> G.add_node("spam") # adds node "spam"
>>> G.add_nodes_from("spam") # adds 4 nodes: 's', 'p', 'a', 'm'
```

At this stage the graph G consists of 8 nodes and 2 edges, as can be seen by:

```
>>> G.number_of_nodes()
8
>>> G.number_of_edges()
2
```

We can examine them with

```
>>> G.nodes()
['a', 1, 2, 3, 'spam', 'm', 'p', 's']
>>> G.edges()
[(1, 2), (1, 3)]
>>> G.neighbors(1)
[2, 3]
```

Removing nodes or edges has similar syntax to adding:

```
>>> G.remove_nodes_from("spam")
>>> G.nodes()
[1, 2, 3, 'spam']
>>> G.remove_edge(1,3)
```

When creating a graph structure by instantiating one of the graph classes you can specify data in several formats.

```
>>> H=nx.DiGraph(G) # create a DiGraph using the connections
from G
>>> H.edges()
[(1, 2), (2, 1)]
>>> edgelist=[(0,1),(1,2),(2,3)]
>>> H=nx.Graph(edgelist)
```

What to use as nodes and edges?

You might notice that nodes and edges are not specified as NetworkX objects. This leaves you free to use meaningful items as nodes and edges. The most common choices are numbers or strings, but a node can be any hashable object (except None), and an edge can be associated with any object x using G.add_edge(n1,n2,object=x).

As an example, n1 and n2 could be protein objects from the RCSB Protein Data Bank, and x could refer to an XML record of publications detailing experimental observations of their interaction.

We have found this power quite useful, but its abuse can lead to unexpected surprises unless one is familiar with Python. If in doubt, consider using convert_node_labels_to_integers() to obtain a more traditional graph with integer labels.

Accessing edges

In addition to the methods Graph.nodes(), Graph.edges() and Graph.neighbors(), iterator versions (e.g., Graph.edges_iter()) can save you from creating large lists when you are just going to iterate through them anyway.

Fast direct access to the graph data structure is also possible using subscript notation.

Warning

Do not change the returned dict–it is part of the graph data structure and direct manipulation may leave the graph in an inconsistent state.

```
>>> G[1] # Warning: do not change the resulting dict
{2: {}}
>>> G[1][2]
{}
```

You can safely set the attributes of an edge using subscript notation if the edge already exists.

```
>>> G.add_edge(1,3)
>>> G[1][3]['color']='blue'
```

Fast examination of all edges is achieved using adjacency iterators. Note that for undirected graphs this actually looks at each edge twice.

```
>>> FG=nx.Graph()
>>> FG.add_weighted_edges_from([(1,2,0.125),(1,3,0.75),(2,4,1.2)
,(3,4,0.375)])
>>> for n,nbrs in FG.adjacency_iter():
... for nbr,eattr in nbrs.items():
... data=eattr['weight']
... if data<0.5: print('(%d, %d, %.3f)' % (n,nbr,data))
(1, 2, 0.125)
(2, 1, 0.125)
(3, 4, 0.375)
(4, 3, 0.375)
```

Convenient access to all edges is achieved with the edges method.

```
>>> for (u,v,d) in FG.edges(data='weight'):
... if d<0.5: print('(%d, %d, %.3f)'%(n,nbr,d))
(1, 2, 0.125)
(3, 4, 0.375)
```

Adding attributes to graphs, nodes and edges. Attributes such as weights, labels, colours or whatever Python object you like, can be attached to graphs, nodes or edges. Each graph, node and edge can hold key/value attribute pairs in an associated attribute dictionary (the keys must be hashable). By default, these are empty, but attributes can be added or changed using add_edge, add_node or direct manipulation of the attribute dictionaries named G.graph, G.node and G.edge for a graph G.

Graph attributes

Assign graph attributes when creating a new graph,

```
>>> G = nx.Graph(day="Friday")
>>> G.graph
{'day': 'Friday'}
```

Or you can modify attributes later

```
>>> G.graph['day']='Monday'
>>> G.graph
{'day': 'Monday'}
```

Node attributes

Add node attributes using add_node(), add_nodes_from() or G.node

```
>>> G.add_node(1, time='5pm')
>>> G.add_nodes_from([3], time='2pm')
>>> G.node[1]
{'time': '5pm'}
>>> G.node[1]['room'] = 714
>>> G.nodes(data=True)
[(1, {'room': 714, 'time': '5pm'}), (3, {'time': '2pm'})]
```

Note that adding a node to G.node does not add it to the graph, use G.add_node() to add new nodes.

Edge attributes

Add edge attributes using add_edge(), add_edges_from(), subscript notation or G.edge.

```
>>> G.add_edge(1, 2, weight=4.7)
>>> G.add_edges_from([(3,4),(4,5)], color='red')
>>> G.add_edges_from([(1,2, {'color':'blue'}),
(2,3,{'weight':8})])
>>> G [1, 2] ['weight'] = 4.7
>>> G.edge[1, 2]['weight'] = 4
```

The special attribute 'weight' should be numeric and holds values used by algorithms requiring weighted edges.

Directed graphs

The DiGraph class provides additional methods specific to directed edges, for example, DiGraph.out_edges (), DiGraph.in_degree (), Digraph. Predecessors (), Digraph. Successors () etc. To allow algorithms to work with both classes easily, the directed versions of neighbors() and degree() are equivalent to successors() and the sum of in_degree() and out_degree(), respectively, even though that may feel inconsistent at times.

```
>>> DG=nx.DiGraph()
>>> DG.add_weighted_edges_from([(1,2,0.5), (3,1,0.75)])
>>> DG.out_degree(1,weight='weight')
0.5
>>> DG.degree(1,weight='weight')
1.25
>>> DG.successors(1)
[2]
>>> DG.neighbors(1)
[2]
```

Some algorithms work only for directed graphs and others are not well defined for directed graphs. Indeed, the tendency to lump directed and undirected graphs together is dangerous. If you want to treat a directed graph as undirected for some measurement you should probably convert it using Graph.to_undirected() or with

```
>>> H = nx.Graph(G) # convert G to undirected graph
```

Multigraphs

NetworkX provides classes for graphs that allow multiple edges between any pair of nodes. The MultiGraph and MultiDiGraph classes allow you to add the same edge twice, possibly with different edge data. This can be powerful for some applications, but many algorithms are not well defined on such graphs. Shortest path is one example. Where results are well defined, for example, MultiGraph.degree(), we provide the function. Otherwise you should convert to a standard graph in a way that makes the measurement well defined [12].

```
>>> MG=nx.MultiGraph()
>>> MG.add_weighted_edges_from([(1,2,.5), (1,2,.75), (2,3,.5)])
>>> MG.degree(weight='weight')
{1: 1.25, 2: 1.75, 3: 0.5}
>>> GG=nx.Graph()
>>> for n,nbrs in MG.adjacency_iter():
... for nbr,edict in nbrs.items():
... minvalue=min([d['weight'] for d in edict.values()])
... GG.add_edge(n,nbr, weight = minvalue)
...
>>> nx.shortest_path(GG,1,3)
[1-3]
```

Graph generators and graph operations

In addition to constructing graphs node-by-node or edge-by-edge, they can also be generated by applying classic graph operations, such as:

1. subgraph(G, nbunch) – induce subgraph of G on nodes in nbunch
2. union(G1,G2) – graph union
3. disjoint_union(G1,G2) – graph union assuming all nodes are different
4. cartesian_product(G1,G2) – return Cartesian product graph
5. compose(G1,G2) – combine graphs identifying nodes common to both
6. complement(G) – graph complement
7. create_empty_copy(G) – return an empty copy of the same graph class
8. convert_to_undirected(G) – return an undirected representation of G
9. convert_to_directed(G) – return a directed representation of G

Using a call to one of the classic small graphs:

```
>>> petersen=nx.petersen_graph()
>>> tutte=nx.tutte_graph()
>>> maze=nx.sedgewick_maze_graph()
>>> tet=nx.tetrahedral_graph()
```

Using a (constructive) generator for a classic graph:

```
>>> K_5=nx.complete_graph(5)
>>> K_3_5=nx.complete_bipartite_graph(3,5)
>>> barbell=nx.barbell_graph(10,10)
>>> lollipop=nx.lollipop_graph(10,20)
```

Using a stochastic graph generator:

```
>>> er=nx.erdos_renyi_graph(100,0.15)
>>> ws=nx.watts_strogatz_graph(30,3,0.1)
>>> ba=nx.barabasi_albert_graph(100,5)
>>> red=nx.random_lobster(100,0.9,0.9)
```

Reading a graph stored in a file using common graph formats, such as edge lists, adjacency lists, GML, GraphML, pickle, LEDA and others:

```
>>> nx.write_gml(red,"path.to.file")
>>> mygraph=nx.read_gml("path.to.file")
```

Details on graph formats: reading and writing graphs

Details on graph generator functions: graph generators

Analyzing graphs

The structure of G can be analyzed using various graph-theoretic functions such as:

```
>>> G=nx.Graph()
>>> G.add_edges_from([(1,2),(1,3)])
>>> G.add_node("spam") # adds node "spam"
>>> nx.connected_components(G)
[[1-3], ['spam']]
>>> sorted(nx.degree(G).values())
[0, 1, 1, 2]
>>> nx.clustering(G)
{1: 0.0, 2: 0.0, 3: 0.0, 'spam': 0.0}
```

Functions that return node properties return dictionaries keyed by node label:

```
>>> nx.degree(G)
{1: 2, 2: 1, 3: 1, 'spam': 0}
```

For values of specific nodes, you can provide a single node or an nbunch of nodes as argument. If a single node is specified, then a single value is returned. If an nbunch is specified, then the function will return a dictionary.

```
>>> nx.degree(G,1)
2
>>> G.degree(1)
2
>>> G.degree([1, 2])
{1: 2, 2: 1}
>>> sorted(G.degree([1, 2]).values())
[1, 2]
>>> sorted(G.degree().values())
[0, 1, 1, 2]
```

Details on graph algorithms supported: algorithms

Drawing graphs

NetworkX is not primarily a graph drawing package but basic drawing with Matplotlib as well as an interface to use the open source Graphviz [13] software package are included. These are part of the networkx.drawing package and will be imported if possible. See Drawing for details.

Note that the drawing package in NetworkX is not yet compatible with Python versions 3.0 and above.

First import Matplotlib's plot interface (Pylab works too)

```
>>> import matplotlib.pyplot as plt
```

You may find it useful to interactively test code using "ipython-pylab", which combines the power of IPython and Matplotlib and provides a convenient interactive mode.

To test if the import of networkx.drawing was successful, draw G using one of:

```
>>> nx.draw(G)
>>> nx.draw_random(G)
>>> nx.draw_circular(G)
>>> nx.draw_spectral(G)
```

when drawing to an interactive display. Note that you may need to issue a Matplotlib

```
>>> plt.show()
```

command if you are not using Matplotlib in interactive mode: (see Matplotlib FAQ)

To save drawings to a file, use, for example:

```
>>> nx.draw(G)
>>> plt.savefig("path.png")
```

writes to the file "path.png" in the local directory. If Graphviz and PyGraphviz, or Pydot, are available on your system, you can also use:

```
>>> nx.draw_graphviz(G)
>>> nx.write_dot(G,'file.dot')
```

Q4. Let's consider a real-world data set: Facebook Ego Networks!

Data Set: https://snap.stanford.edu/data/egonets-Facebook.html

This Facebook-combined ego networks data set contains the aggregated network of ten individuals' Facebook friends list. In this data set, the vertices represent individuals on Facebook, and an edge between two users means they are Facebook friends. Each ego network was created by starting with a Facebook user and then mining all of his Facebook friends. The ego networks are then aggregated by identifying individuals who appear in multiple ego networks [14, 15].

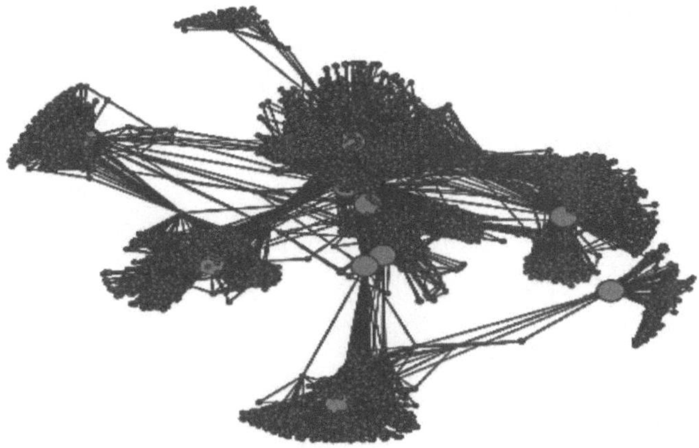

FIGURE 12.19 Parallel betweenness [16]

From this data set, solve the following problems of social network analysis using the Python library NetworkX and also visualize these two statistics using the Gephi tool.

Parallel Betweenness: How do we determine who is the most 'important' individual in the network. Well, that depends on the definition of 'important'. One way to define 'importance' is the individual's betweenness centrality. The betweenness centrality is a measure of how many shortest paths pass through a particular vertex. The shortest paths that pass through the vertex, the more central the vertex is to the network [16].

FIGURE 12.20 Graph after community detection [16]

Community Detection: Facebook friends probably come from different aspects of your life; some are your friends from college, others are your co-workers and maybe some old friends from your hometown. Because your friends can be broken down into different groups like this, you may wonder if we can identify these different *communities* in your social network. Using *community detection* algorithms, we can break down a social network into different potentially overlapping communities [16].

REFERENCES

1. Barnes, J. A. Class and Committees in a Norwegian Island Parish. *Human Relations* 7:39–58
2. Borgatti, S. (2002). *NetDraw Software for Network Visualization*. Lexington: Analytic Technologies.
3. Borgatti, S. E. (2002). *Ucinet for Windows: Software for Social Network Analysis.* Harvard: Analytic Technologies.
4. Berkowitz, S. D. 1982. *An Introduction to Structural Analysis: The Network Approach to Social Research*. Toronto: Butterworth.
5. Gephi. https://gephi.org/
6. Wassink, A. (2018). UCINet quick guide. https://depts.washington.edu/sociolab/researchtools/networkanalysis/documents/UCINetQuickGuide2018.pdf
7. Social Media Research Foundation. Tutorials. https://www.smrfoundation.org/nodexl/tutorials/
8. Kabacoff, R. I. (2017). R tutorial. https://www.statmethods.net/r-tutorial/index.html
9. *NetLogo 6.2.0 User Manual*. Tutorial #1: Models. https://ccl.northwestern.edu/netlogo/docs/tutorial1.html
10. Batagelj, V., & Mrver, A. (2012, Sep. 24). *Pajek Reference Manual*. http://vlado.fmf.uni-lj.si/pub/networks/pajek/doc/pajekman.pdf
11. NetworkX tutorial: Creating a graph. https://networkx.org/documentation/stable/tutorial.html
12. NetworkX tutorial: Creating a graph. https://networkx.readthedocs.io/en/v1.10/tutorial/tutorial.html
13. Social Network Visualizer. https://socnetv.org/
14. Kumar, V. (2021, Aug. 3). 23 free social network analysis tools [as of 2021]. *RankRed*. https://www.rankred.com/free-social-network-analysis-tools/
15. Wikipedia. Social network analysis software. https://en.wikipedia.org/wiki/Social_network_analysis_software
16. Nandi, M. (2015, July 14). Social network analysis with NetworkX. *Domino Data Lab*. https://blog.dominodatalab.com/social-network-analysis-with-networkx/

Index

Q

R

S

T

U

V

W

Y

Z

FSC
www.fsc.org
MIX
Papier | Fördert gute Waldnutzung
FSC® C083411

Kolibri 360 GmbH
Ferdinand-Jühlke-Straße 7
99095 Erfurt, Deutschland
produktsicherheit@kolibri360.de